TREES OF LIFE

AUSTRALASIAN STUDIES IN HISTORY AND PHILOSOPHY OF SCIENCE

VOLUME 11

The titles published in this series are listed at the end of this volume.

TREES OF LIFE

Essays in Philosophy of Biology

Edited by

PAUL GRIFFITHS

University of Otago,
Dunedin, New Zealand

KLUWER ACADEMIC PUBLISHERS

DORDRECHT / BOSTON / LONDON

Library of Congress Cataloging-in-Publication Data

Trees of life : essays in philosophy of biology / edited by Paul Griffiths.
p. cm. -- (Australasian studies in history and philosophy of science ; v. 11)
Includes bibliographical references and index.

1. Evolution (Biology)--Philosophy--Congresses. 2. Biology--Philosophy--Congresses. I. Griffiths, Paul. II. Series.
QH371.T72 1992
575'.001--dc20 92-8364

ISBN 978-90-481-4153-1

Published by Kluwer Academic Publishers,
P.O. Box 17, 3300 AA Dordrecht, The Netherlands.

Kluwer Academic Publishers incorporates
the publishing programmes of
D. Reidel, Martinus Nijhoff, Dr W. Junk and MTP Press.

Sold and distributed in the U.S.A. and Canada
by Kluwer Academic Publishers,
101 Philip Drive, Norwell, MA 02061, U.S.A.

In all other countries, sold and distributed
by Kluwer Academic Publishers Group,
P.O. Box 322, 3300 AH Dordrecht, The Netherlands.

Printed on acid-free paper

Printed in the Netherlands

FOREWORD

Australia and New Zealand boast an active community of scholars working in the field of history, philosophy and social studies of science. 'Australasian Studies in History and Philosophy of Science' aims to provide a distinctive publication outlet for their work. Each volume comprises a group of essays on a connected theme, edited by an Australian or a New Zealander with special expertise in that particular area. In each volume, a majority of the contributors is from Australia or New Zealand. Contributions from elsewhere are by no means ruled out, however, and are indeed actively encouraged wherever appropriate to the balance of the volume in question. Earlier volumes in the series have been welcomed for significantly advancing the discussion of the topics they have dealt with. The present volume will I believe be greeted equally enthusiastically by readers in many parts of the world.

R. W. Home
General Editor
Australasian Studies in History and Philosophy of Science

TABLE OF CONTENTS

Paul Griffiths

INTRODUCTION*

The papers in this volume are a selection of those presented at the Philosophical Problems in Evolutionary Biology conference, held at the University of Otago, Dunedin, New Zealand in August 1990. The occasion for this conference was a visit to Otago by Elliott Sober. Sober's work has had a major influence on the philosophy of biology, and his visit was an ideal opportunity to bring together interested philosophers and biologists from all over New Zealand. Having organised a conference to coincide with Sober's visit, we were fortunate enough to attract a number of other overseas speakers, some of whose papers appear in this volume, together with Sober's and those of local speakers.

The volume is divided into three sections. The first, 'Beyond Natural Selection' is concerned with attempts to extend the forms of explanation found in evolutionary biology to new domains. Elliott Sober's paper 'Models of Cultural Evolution' examines some recent attempts to give 'evolutionary' models of cultural change. Sober is not concerned to discuss the well-known sociobiological models of writers such as E.O. Wilson[1]. These models attempt to explain social phenomena as the results of straightforward natural selection, and have been widely discussed elsewhere[2]. Instead, Sober draws attention to models which make use of modes of transmission other than genetic inheritance, or which measure fitness by something other than number of offspring. Models which abandon both of these are, as Sober notes, only vaguely analogous to natural selection. Nevertheless, the success of explanation by natural selection is often urged as a reason to take such models seriously.

The models which Sober examines in detail are that of Cavalli-Sforza and Feldman and that of Boyd and Richerson[3]. These theorists

P. Griffiths (ed.), Trees of Life, 1–13.

see the mechanisms of cultural evolution as products of natural selection, but products which have now become forces for change in their own right. Ideas, behaviours and institutions can be transmitted by means other than inheritance, giving rise to patterns different from those encountered in conventional evolution. The 'fitness' of such cultural items may be quite unrelated to the number of descendants of those who adhere to them. These ideas are a useful corrective to traditional sociobiology. Such genuinely independent mechanisms can oppose, reinforce and otherwise interact with conventional evolution to determine actual outcomes. There is no sense in which conventional evolution by change in gene frequencies is more 'basic' than cultural change, nor need genetic evolution always overpower cultural when the two are opposed.

Despite this, Sober is sceptical of the value of these theories. His scepticism is expressed in terms of the distinction between 'source laws' and 'consequence laws' drawn in his influential book 'The Nature of Selection'[4]. In that book, Sober construed evolutionary theory as a theory of interacting forces. Population genetics, the mathematical heart of modern evolutionary theory, provides the 'consequence laws' of this theory. Consequence laws state what actual change will occur as the result of the interaction of a number of forces. Population genetics tells us how the relative fitnesses of traits together with factors such as genetic drift, interact to determine actual changes in gene frequencies. But a theory of forces also requires 'source laws'. These describe the origins of the forces to which consequence laws apply. In biology, various disciplines contribute to our understanding of the actual basis of fitness differences, and thus provide source laws.

The theories of cultural evolution Sober considers are all modelled on population genetics. They try to model change in cultures due to the relative 'fitnesses' of elements of culture, and phenomena such as drift. The social sciences, however, have been more concerned with source laws. They set out to discover what has made particular ideas attractive, or why the poorer classes of a society copy the changing manners of the better off with such and such a time-lag. Our understanding of these source laws is very limited, and it is a matter of dispute whether it can ever be anything more. Sober suggests that quantitative models of social change based on population genetics will seem merely irrelevant to social scientists until they have uncovered the qualitative causes of social change. Without such progress in the conventional social science, it will be impossible to apply the models to particular cultural phenomena.

In 'Punctuated Equilibrium and Macroevolution', Kim Sterelny considers an attempt to extend natural selection in another plane. Here we are still concerned with the history of life, but with a larger view of this history, where the actors are not individual organisms, but larger units such as species. Sterelny distinguishes three theses concerning macroevolution. The 'pattern hypothesis' states that species typically come into being in relatively short periods of evolutionary time, remain stable for a long period, and then go extinct again in a short period. Sterelny takes the view that it is at least an open empirical question whether this pattern exists. The second macroevolutionary thesis, the 'process hypothesis', claims that unconventional evolutionary processes are needed to explain the pattern. It is alleged that the periods of rapid change require saltative leaps from one species form to another, rather than the gradual change of conventional Darwinism. It is further alleged that the stability of species over long periods of time tells against the traditional picture of evolution as gradual adaptation to the environment. If species are always subject to natural selection, then they should be in a permanent state of flux. Sterelny argues that conventional Darwinism has the resources to explain both these phenomena, and that the pattern hypothesis, even if correct, provides little reason to postulate unconventional evolutionary processes.

The third macroevolutionary hypothesis which Sterelny considers is more subtle. It suggests that, although evolution is driven by conventional Darwinian processes at the individual level, there are larger scale patterns in evolution that cannot be discerned at this level. A refusal to consider species as real actors in the evolutionary process will cause us to overlook these patterns. Sterelny accepts that it may be possible to describe evolutionary processes at the species level. Species might be described as replicators, akin to genes, carrying information in their gene pool and replicating with variation in speciation events. Species may also be seen as vehicles – macro-phenotypes for their gene pools – which compete with one another and are selected in virtue of their differing fitnesses. Sterelny allows that both kinds of stories can be made coherent, and discusses some attempts to do so, particularly Schull's[5] theory of 'species-intelligence'. He doubts, however, that there are any such stories that cannot be adequately expressed in more conventional terms, where individual organisms and local populations are the main actors. Sterelny and Kitcher[6] have argued that there can be non-competing alternative descriptions of evolutionary processes at different 'levels of selection'. The more plausible macroevolutionary stories appear to provide just such alternative descriptions of known processes.

Sterelny finds only one example of a what may be a genuine, irreducible macroevolutionary process. This is the process of speciation itself. Speciation, and types of speciation, occur many times in the evolutionary tree. Different speciation events may be recognisably type-identical despite huge differences in the species involved in these events, and in the individual level details of the events. Speciation events may also play an important role in evolutionary theory, explaining, for example, how useful mutations are not swamped by backcrossing. If this is correct, then reducing speciation events to the details of what befell particular organisms and local populations would prevent us from expressing true and interesting generalisations about evolution.

The final paper in this section is Robin Craw's highly original contribution to the programme of 'evolutionary' sociology of science initiated by David Hull's 'Science as a Process'[7]. Craw's 'Margins of Cladistics' reanalyses Hull's own case study – the emergence of cladistics as the dominant approach to systematics. He shows that Hull's single line of descent via Willi Hennig in Germany and Gareth Nelson in New York – reminiscent of a pre-Darwinian chain of being – is just one route through a complex tree of interacting research traditions in many countries. Craw reveals the development of cladistic methods in Italy and the United States independently of Hennig, the influence of Australasian systematists on Hennig's own work, and the spread of Hennig's ideas to Australasia and other parts of the world independently of, and indeed prior to, the publicity given to his work by Nelson.

The central theoretical point of Craw's paper is that there is a spatial as well as a temporal aspect to intellectual evolution. The process cannot be described as a temporal sequence of events, because many events are occurring in different places at the one time. It is possible to see the paper as the application to intellectual history of the strongly biogeographic emphasis in systematics that has become something of a New Zealand speciality in recent years. Craw's 'difference that place makes' is as startling in intellectual as in biological evolution.

Section two contains papers dealing with the central concepts of evolutionary theory, adaptation & selection. Paul Griffiths' 'Adaptive Explanation and the Concept of Vestige' argues that adaptive or functional, explanations were unfairly caricatured in various attacks on 'adaptationism' in the 1980's. Griffiths discusses Gould and Vrba's famous distinction between 'adaptations' and 'exaptations', traits which evolved because they conferred a selective advantage and traits which exist for other reasons and happen accidentally to be useful to the organism. He argues that thinking in these terms hides the real potential of adaptive explanation. Drawing on recent philosophical work on the difference between functions and effects of biological traits[8], he suggest a new, tripartite distinction. Gould and Vrba assumed that once a trait has been acquired, its biological function is determined. If it was not an adaptation for some function at the outset then it must be an exaptation. Griffiths argues instead that a trait can become an adaptation for some function at a later stage in its history ('exadaptation'). This is because traits are not just spread through the population by selection, they are maintained there by selection. Selection does not only explain branching events on cladograms, it can explain the maintenance of traits 'along the branches'. If the selection pressures maintaining a trait changes, then the function of the trait may change too. A trait which was an adaptation for one purpose thus becomes an adaptation for another.

The apparatus Griffiths constructs in an attempt to be fair to adaptationism also allows him to construct a definition of vestigiality. Vestigial traits arose as adaptations but now fail to perform their original function, and perform some other function, or no function at all. Some care is needed in making this notion precise. Vestigial traits are traditionally pictured as reduced or atrophied, but the role played by this feature in the definition of vestigiality is suprising[9].

Timothy Shanahan's 'Selection, Drift and the aims of Evolutionary Theory' is a careful and detailed examination of the distinction between these two fundamental concepts. Shanahan argues that no intrinsic difference can be found between 'sampling events' which are examples of selection and those which are examples of drift. He questions the traditional characterisation of a selection as sampling on the basis of fitness differences, arguing that at the individual level, fitness differences have no actual causal role in the survival and reproduction of organisms. The fate of individuals is decided by the interaction of a subset of their properties with a subset of the important features of the environment. Many, and perhaps all, the properties which lead us to ascribe an organism of that type a given fitness level will play no role in determining the individuals fate. Identifying these properties as components of fitness involves seeing the organism as one of a type, and considering the probabilities of an organism of that type interacting with certain types of environmental features. 'Sampling on the basis of fitness differences' is something that happens to populations, not to individuals. In evolutionary theory, as in statistical thermodynamics, there are properties of arrays that make little or no sense when applied to individual particles.

Shanahan agrees that selection and drift are distinct evolutionary processes, but the distinction is not between two types of event, but between two types of patterns of events. An event is an example of selection because it can be grouped with enough similar events to have a potential for giving rise to adaptations.

The final section of the book contains three papers on the relationship between evolutionary and developmental biology.

Gray's 'Death of the Gene' is a systematic defense of the 'developmental systems' or 'constructionist' approach to evolution. It is a truism that every phenotypic trait requires both genic and environmental inputs. But it is still common to see the genes as carrying information about traits, while the environment merely provides necessary 'support'. There are also techniques which purport to determine by statistical means the relative contribution of the two factors to some trait (approaches famously criticised by Richard Lewontin[10]). Gray argues that none of these approaches are useful. He suggests that the fundamental unit of evolution is the 'developmental system' - a set of organismic and environmental features which produce an outcome capable of replicating the developmental process. Evolution is the differential replication of these systems. Genic and environmental factors are actually abstractions from this larger system, as genes only produce a determinate outcome in a given developmental context, and environmental factors only have a determinate effect given an organismic system to act upon.

One factor stressed in Gray's work is extra-genomic inheritance. Developmental systems do not just inherit their genomes, they inherit many other environmental features necessary for normal developmental outcomes. This is particularly clear when such inheritance involves the active re-creation of the environmental feature by the population. Structured social or physical environments are often necessary for normal developmental outcomes. These extra-organismic features evolve along with the more narrowly conceived phenotype, and are passed on from generation to generation. Gray points out that changes in these extra-genic features can cause a changes in the evolutionary trajectory of a lineage of developmental systems that is just as stable and self-perpetuating as genetic change.

The New Zealand biogeographic tradition asserts itself again in Gray's treatment of the inheritance of geographic features. Many features which are vital parts of developmental systems persist across generations without active replication. Although these features are not differentially perpetuated, Gray suggests that their history, and the history of their interaction with other features of developmental systems, is a part of evolution. As Leon Croizat's famous slogan has it, 'Earth and life evolve together'.

Gray relates his approach to recent work in the units of selection debate. He takes Sober's[11] arguments for a organismic, rather than genic, approach to evolution and suggests that their logic can be extended to favour the new, still larger unit. He also produces an extended rejoinder to Kim Sterelny and Philip Kitcher's 1988 paper 'The Return of the Gene'[12]. This paper is itself a response to the work of Richard Lewontin, Elliott Sober and others on the implications of developmental theory[13]. These workers draw attention to the fact that one gene may be implicated in the control of many traits (pleitropy) and that the effects of a gene depend on the rest of the genome and on the extra-genic environment (epigenesis). As the environment, especially the genic environment, is subject to frequent change, they suggest that it is difficult to think of particular genes as 'controlling for' particular phenotypic features. They argue further that because genes make a varying contribution to the phenotype, they are not stable positive or negative factors in organismic fitness and cannot properly be regarded as targets of selection. Sterelny and Kitcher show that by a delicate system of indexing, it is possible to construct a precise sense in which loci on the genome can be said to control for particular ranges of traits, and alleles or allelic pairs to control for particular values of that trait. They argue that the existence of standard environments and common alternative alleles make the contribution of a gene stable enough to allow genes to be targets of selection.

Gray is unimpressed by this attempt to revive the gene. The only notion of 'standard environment' robust enough to support Sterelny and Kitcher's claims is that of 'intended environment' - the environment which played a role in the evolutionary history of the lineage, and which produces developmental outcomes of the sort that allowed this lineage to proliferate. A standard environment, in other words, is the rest of the developmental system. But these extra-genomic features of the system are themselves the product of evolution. By backgrounding these features, the genic selectionist eschews part of the explanatory potential of evolutionary theory. To take Gray's eucalypt example, the gene selectionist offers an explanation of the evolution of genes for fire-germinated gumnuts and inflammable detritus. Developmental systems theory offers in addition a theory of the evolution of bushfires, and of the interaction of fire and tree.

One exciting feature of Gray's treatment of the units of selection debate is a threat to the claims of genic selectionists to be able to represent every genuine evolutionary process. Through the use of extra-genic inheritance, Gray allows for cases where a lineage of developmental systems incorporates a new developmental feature, which alters the developmental outcome and allows the lineage to out compete its rivals, all with no change in gene frequency. An example might occur in a species which exhibits variations in a culturally inherited mate recognition system in various sympatric lineages. If one such recognition system confers some other advantage on a lineage, then there will be a piece of evolutionary history that cannot be explained in genic terms. This possibility gives the developmental systems approach a unique card in the units of selection game. All other alternatives to genic selectionism have claimed to give a better picture of the real causal forces in selection, at the expense of limited scope. Developmental systems theory may be able to claim maximal generality as well as faithfulness to the actual process.

Susan Oyama has played an important role in the growth of the 'developmental systems' approach, especially through her book 'The Ontogeny of Information'[14]. In 'Ontogeny & Phylogeny' Oyama compares the debate over the relative importance of phylogenetic constraint and natural selection in evolution, a debate which exercised many important theorists in the seventies and eighties, to the older debate over the role of genetic and environmental factors ('nature' and 'nurture') in development. In both cases, the outcome of a process is conceived as the result of two sets of interacting forces, internal forces, due to the constitution of the system which is evolving, and external forces, due to the environment in which it evolves. The developmental systems approach has already rejected this dichotomy in the case of individual development (ontogeny). Oyama suggests that it should also be rejected in the case of evolutionary development (phylogeny).

The developmental constraints which have been suggested in recent literature as major factors in phylogeny may seem to obey the developmental systems theorists injunction to integrate developmental biology into evolution. The proponent of developmental constraints claims that evolution cannot select advantageous genes at will, but only combinations which will fit together into a viable developmental system. This places interesting and principled constraints on the possible deviations from standard forms, and helps shape the general path of evolution. But Oyama suggests that because these postulated constraints exist in the organism independently of its environment, they are untenable in the same way as the traditional genetic programme. The 'developmental system' as conceived by the proponent of developmental constraints is a system with fixed potentialities. Oyama's developmental system, on the other hand, consists of a genome-in-environment. The potential of the system for new developments cannot be specified independently of the future environments which may come to be. If there are developmental systems which are not viable, this is not because of any pre-existing constraint or tendency in the 'basic plan'[15] of the broader taxon which is evolving.

In the final paper in the collection, 'Against Ontogeny', John Morss takes issue with the idea that the development of an individual organism proceeds in an orderly fashion through a series of predetermined stages - the traditional conception of ontogeny. This notion has had an enormous influence on developmental psychology, as documented in Morss' book 'The Biologising of Childhood'[16]. The first part of Morss' paper documents the introduction and growth of the idea of ontogeny in developmental theory. The idea was introduced via the now discredited Haeckelian notion that 'ontogeny recapitulates phylogeny'. The development of the individual organism was imagined to be a recapitulation of evolutionary history, displaying the sequence of ancestral forms. Furthermore, the phylogeny that was supposed to be recapitulated was a progressivist phylogeny, in which each stage of evolution represented a step up the chain of being! Morss argues that the current, ontogenetic, view of psychological development exists as a result of this tradition, and must be justified afresh in terms of current evolutionary thinking.

The second part of Morss' paper looks at ontogeny in recent evolutionary theory. A number of contemporary theorists have criticised linear models of ontogeny, in which the organism is seen as passing through a single sequence of stages. The most popular alternative is a 'developmental pathways' model. In this more flexible conception of ontogeny, the organisms development can take a number of paths, as determined by interactions with the environment at critical points in development.

Morss admits that the developmental pathways model meets many of his concerns, but concludes the paper with a more foundational objection to the whole idea of ontogeny. His objection is based on the now famous idea of David Hull and Michael Ghiselin[17] that species are not natural kinds, but rather historical individuals. Morss notes that this view, now widely accepted, implies that individuals are not subject to genuine laws of nature qua members of a given species. Species are collections of more or less similar individuals, associated by their common history. There will be true generalisations about the members of a species, in virtue of their similarity and common antecedents, but these generalisations will not be lawlike, any more than the generalisation that the products of the Citröen car company are beautiful but costly to maintain. From this theoretical perspective there can be no 'laws of development' for a species. The most there can be is a pattern to which typical members of the species more or less roughly conform because of the similarities in their genomes and environment.

LIST OF AUTHORS

Elliott Sober is Professor of Philosophy at the University of Wisconsin at Madison, Wisconsin, USA.

Kim Sterelny is Senior Lecturer in Philosophy at Victoria University of Wellington, Wellington, New Zealand.

Robin Craw is Systematist at the Department of Scientific and Industrial Research Plant Protection Unit, Mt. Albert Research Centre, Auckland, New Zealand.

Paul Griffiths is Lecturer in Philosophy at the University of Otago, Dunedin, New Zealand.

Timothy Shanahan is Assistant Professor of Philosophy at Loyala Marymount University, Los Angeles, California, USA.

Russell Gray is Lecturer in Psychology at the University of Otago, Dunedin, New Zealand.

Susan Oyama is Professor of Psychology at the City University of New York, New York, USA.

John Morss is Senior Lecturer in Education at the University of Otago, Dunedin, New Zealand.

NOTES

* I would like to acknowledge the financial assistance of the Department of Philosophy, Division of Humanities and Division of Sciences of the University of Otago, without which this volume could not have been prepared, and the work of my research assistant, Michael Thrush, who is responsible for its physical appearance.

[1] Wilson, E.O. *On Human Nature* Harvard University Press 1978; Wilson, E.O & Lumsden, C.J. *Genes, Minds & Culture* Harvard University Press 1981.

[2] Kitcher, P. *Vaulting Ambition* MIT Press, Camb. Mass. 1985.

[3] L. Cavalli Sforza & M . Feldman (1981), *Cultural Transmission & Evolution: A Quantitative Approach* , Princeton University Press. Boyd, R. & Richerson, P. (1985), *Culture & the Evolutionary Process* , University of Chicago Press.

[4] Sober, E. *The Nature of Selection* . Bradford Books/MIT Press. Cambridge, Mass. 1984.

[5] Schull 'Are Species Intelligent?' *Behavioural & Brain Sciences* , 1990 p63-108.

[6] Sterelny, K & Kitcher, P. The Return of the Gene. *Journal of Philosophy,* **LXXXV**. 1988 p339-361.

[7] Hull, D. *Science as a Process.* University of Chicago Press. 1988.

[8] Neander, K. 'Functions as Selected Effects', *Philosophy of Science* (Forthcoming); Griffiths, P.E. 'Functional Analysis and Proper Functions', *British Journal for Philosophy of Science* (Forthcoming).

[9] Since the submission of this paper for the volume, an interesting

paper has appeared by David Baum & Allan Larson on very similar topics. ('Adaptation reviewed: A Phylogenetic methodology for Studying Character Macroevolution', *Systematic Zoology* **40(1)** 1991. p1-18). Baum & Larson are dissatisfied with Gould and Vrba's concepts for similar reasons to Griffiths. Their revised taxonomy, however, is very different, being driven by the need to generate hypotheses testable by cladistic methods, rather than considerations from the theory of biological teleology. It would be an interesting, though substantial task, to determine how far the two approaches are compatible.

[10] Lewontin, R. 'The Analysis of Variance and the Analysis of Causes', *American Journal of Human Genetics*. **26**. 1974. p400-411.

[11] Sober, E. Op.cit., & Sober, E & Lewontin, R. 'Artifact, Cause & Genic Selection', *Philosophy of Science*. 1982. p157-180.

[12] Op.cit.

[13] Lewontin, R. Op.cit.; 'Gene, Organism & Environment'. In *Evolution: From Molecules to Man*. CUP. 1983. p273-285; and especially Sober, E & Lewontin, R. Op. cit. Oyama, S. *The Ontogeny of Information: Developmental Systems and Evolution*, Cambridge University Press. 1985.

[14] Oyama, S. *The Ontogeny of Information: Developmental Systems and Evolution*, Cambridge University Press. 1985.

[15] See Gould, S.J. & Lewontin, R. 'The Spandrels of San Marco and the Panglossian Paradigm: A Critique of the Adaptationist Programme' Reprinted in Sober, E (Ed) *Conceptual Problems in Evolutionary Biology*. MIT Press, Cambridge, Mass. 1984.

[16] Morss, J.R. *The Biologising of Childhood*. Lawrence Erlbaum Associates. 1990.

[17] Hull, D. 'Are Species Really Individuals' *Systematic Zoology* **25** 1976. 174 - 191; Ghiselin, M. 'A Radical Solution to the Species Problem'. *Systematic Zoology*. **23**. 1975. 536-544.

SECTION ONE

BEYOND NATURAL SELECTION

Elliott Sober

MODELS OF CULTURAL EVOLUTION[1]

At least since the time of Darwin, there has been a tradition of borrowing between evolutionary theory and the social sciences. Darwin himself owed a debt to the Scottish economists who showed him how order can be produced without conscious design. Adam Smith thought that socially beneficial characteristics can emerge in a society as if by an "invisible hand;" though each individual acts only in his or her narrow self-interest, the result, Smith thought, would be a society of order, harmony, and prosperity. The kind of theory Darwin aimed at — in which fitness improves in a population without any conscious guidance — found a suggestive precedent in the social sciences.

The use of game theory by Maynard Smith[2] and others provides a contemporary example in which an idea invented in the social sciences finds application in evolutionary theory. Economists and mathematicians were the first to investigate the pay-offs that would accrue to players following different strategies in games of a given structure. Biologists were able to see that game theory does not require that the players be rational nor even that they have minds. The behaviour of organisms exhibits regularities; this is enough for us to talk of them as pursuing strategies. The pay-offs of the behaviours that result from these strategies can be measured in the currency of fitness — i.e., in terms of their consequences for survival and reproduction. This means that the idea of pay-offs within games allows us to describe evolution by natural selection. Here again is a case in which a social scientific idea has broader scope than its initial social science applications might have suggested.

At present, there is considerable interest and controversy sur-

P. Griffiths (ed.), Trees of Life, 17–39.

rounding borrowings that go in the opposite direction. Rather than apply social science ideas to biological phenomena, sociobiology and related research programs aim to apply evolutionary ideas to problems that have traditionally been thought to be part of the subject matter of the social sciences. Sociobiology is the best known of these enterprises. It has been criticized on a variety of fronts. Although I think that these criticisms differ in their force, I don't want to review them here. My interest is in a somewhat lesser known movement within biology, one that strives to extend evolutionary ideas to social scientific phenomena. I want to discuss the models of cultural evolution put forward by Cavalli-Sforza and Feldman[3] and by Boyd and Richerson.[4] These authors have distanced themselves from the mistakes they see attaching to sociobiology. In particular, they wish to describe how cultural traits can evolve for reasons that have nothing to do with the consequences the traits have for reproductive fitness. In a very real sense, their models describe how it is possible for mind and culture to play an irreducible and autonomous role in cultural change. For this reason, there is at least one standard criticism of sociobiology that does not apply to these models of cultural evolution. They deserve a separate hearing.

In order to clarify how these models differ from some of the ideas put forward in sociobiology, it will be useful to describe some simple ways in which models of natural selection can differ. I focus here on natural selection, even though there is more to evolutionary theory than the theory of natural selection, and in spite of the fact that the two books I am considering sometimes exploit these nonselectionist ideas. Although there are nonselectionist ideas in these two books, the bulk of their models assign a preeminent role to natural selection and its cultural analogs. So a taxonomy of selection models will help us see how models of cultural evolution are related to arguments put forward in sociobiology.

There are two crucial ingredients in a selection process. Given a set of objects that exhibit variation, what will it take for that ensemble to evolve by natural selection? By evolution, I mean that the frequency of some characteristic in the population changes. The first requirement is that the objects differ with respect to some characteristic that makes a difference in their abilities to survive and reproduce. Secondly, there must be some way to insure that offspring resemble their parents. The first of these ingredients is called *differential fitness*; the second is *heritability*.

In standard formulations of the genetical theory of natural selection, different genes or gene complexes in a population encode different phenotypes. The phenotypes confer different capacities to survive and reproduce on the organisms that possess them. As a result, some genes are more successful in finding their way into the next generation than others. In consequence, the frequency of the phenotype in question changes. This is evolution by natural selection with a genetic mode of transmission. Note that traits differ in fitness because some organisms have more babies than others. It may seem odd to say that "having babies"[5] is one way to measure fitness, as if there could be others. My reason for saying this will become clearer later on.

The phenotype treated in such a selection model might be virtually any piece of morphology, physiology, or behaviour. Biologists have developed different applications of this Darwinian pattern to characteristics of all three sorts in a variety of species. One way — the most straightforward way — to apply biology to the human sciences is to claim that some psychological or cultural characteristic became common in our species by a selection process of this sort. This is essentially the pattern of explanation that Wilson was using when he talked about aggression, xenophobia, and behavioural differences between the sexes. An ancestral population is postulated in which phenotypic differences have a genetic basis; then a claim is made about the consequences of those phenotypes for survival and reproduction. This is used to explain why the population changed to the configuration we now observe.

The second form that a selection process can take retains the idea that fitness is measured by how many babies an organism produces, but drops the idea that the relevant phenotypes are genetically transmitted. Strictly speaking, evolution by natural selection does not require genes. It simply requires that offspring resemble their parents. For example, if characteristics were transmitted by parents teaching their children, a selection process could occur without the mediation of genes.

A hypothetical example of how this might happen is afforded by that favourite subject of sociobiological speculation — the incest taboo. Suppose that incest avoidance is advantageous because individuals with the trait have more viable offspring than individuals without it. The reason is that outbreeding diminishes the chance that children will have deleterious recessive genes in double dose. If offspring learn whether to be incest avoiders from their parents, the frequency of the trait in the population may evolve. And this may occur without there being any genetic differences between those who avoid incest and those who do not. Indeed, incest avoidance could evolve in this way in a population of genetically identical individuals, provided that the environmental determinant of the behaviour runs in families.6

In this second kind of selection model, mind and culture displace one but not the other of the ingredients found in models of the first type. In the first sort of model, a genetic mode of transmission works side-by-side with a concept of fitness defined in terms of reproductive output — what I have called "having babies." In the second, reproductive output is retained as the measure of fitness, but the genetic mode of transmission is replaced by a psychological one. Teaching can provide the requisite heritability just as much as genes.

The third pattern for applying the idea of natural selection abandons both of the ingredients present in the first. Genes are abandoned as the mode of transmission. And fitness is not measured by how many babies an organism has. Individuals acquire their ideas because they are exposed to the ideas of their parents, of their peers, and of their parents' generation. So the transmission patterns may be vertical, horizontal, and oblique. An individual exposed to a mix of ideas drawn from these different sources need not give them all equal credence. Some may be more attractive than others. If so, the frequency of ideas in the population may evolve over time. Notice that there is no need for organisms to differ in terms of their survivorship or degree of reproductive success in this case. Some ideas catch on while others become passe'. In this third sort of selection model, ideas spread the way a contagion spreads.

It is evident that this way of modelling cultural change is tied to the genetical theory of natural selection no more than it is tied to epidemiology. Rumours and diseases exhibit a similar dynamic. The spread of a novel characteristic in a population by natural selection, like the spread of an infection or an idea, is a diffusion process.

This third type of selection model has a history that predates sociobiology and the models of cultural evolution that I eventually want to discuss. Consider the economic theory of the firm.[7] Suppose one wishes to explain why businesses of a certain sort in an economy behave as profit maximizers. One hypothesis might be that individual managers are rational and economically well informed; they adjust their behaviour so as to cope with market conditions. Call this the learning hypothesis. An alternative hypothesis is that managers are not especially rational or well informed. Rather, firms that are not efficient profit maximizers go bankrupt and thereby disappear from the market. This second hypothesis posits a selection process.

Note that the selection hypothesis involved here is of type III. Individual firms stick to the same market strategies, or convert to new ones, by some process other than genetic transmission. In addition, the biological kind of survival and reproduction (what I have called "having babies") does not play a role. Firms survive differentially, but this does not require any individual organism to die or reproduce.

A different example of type III models, which will be familiar to philosophers of science, is involved in some versions of evolutionary epistemology. Karl Popper suggested that scientific theories compete with each other in a struggle for existence.[8] Better theories spread through the population of inquirers; inferior ones exit from the scene. Popper highlighted the nonbiological definition of fitness used in this view of the scientific process when he said that "our theories die in our stead."[9]

The three possible forms that a selection model can take are summarized in table 1. By "learning," I don't want to require anything that is especially cognitive; imitation is a kind of learning. In addition, "having students" should be interpreted broadly, as any sort of successful influence mediated by learning.[10]

The parallelism between type I and type III models is instructive. In the type I case, individuals produce different numbers of babies in virtue of their phenotypic differences (which are transmitted genetically). In the type III case, individuals produce different numbers of students in virtue of their phenotypic differences (which are

	Heritability	Fitness
I	Genes	Having Babies
II	Learning	Having Babies
III	Learning	Having Students

Table 1. Three Types of Selection Model[11]

transmitted by learning).

Selection models of cultural characteristics that are of either pattern I or pattern II can properly be said to provide a "biological" treatment of the characteristic in question. Models of type III, on the other hand, do not really propose biological explanations at all. A selectional theory of the firm, or a diffusion model that describes the spread in popularity of an idiom in a language, are no more "biological" than their competitors. In type III models, the mode of transmission and the reason for differential survival and replication may have an entirely autonomous cultural basis. Genes and having babies are notable by their absence; the biological concept of natural selection plays the role of a suggestive metaphor, and nothing more.

It is important to recognize that this three-fold taxonomy describes the process of natural selection, not the product that process may yield. For example, once a type I process of natural selection has run its course, it is an open question whether the variation that remains is genetic or nongenetic. Consider the work in sociobiology by Richard Alexander.[12] He believes that human beings behave so as to maximize their inclusive fitness. This means that there is an evolutionary explanation for the fact that people in one culture behave differently from those in another. But Alexander does not think that this is due to there being genetic differences between the two cultures. Rather, his idea is that the human genome has evolved so that a person will select the fittest behaviour, given the environment he or she occupies. The fact that people behave differently is due to the fact that they occupy different environments. So, in terms of the current variation that we observe, Alexander is, in fact, a radical environmentalist. This is worth contemplating if you think that sociobiology stands or falls with the thesis of genetic determinism.

Matters change when we consider not the present situation, but the evolutionary past that generated it. The genome that Alexander postulates, which gives current humans their ability to modify behaviour in the light of ecological conditions, evolved because it was fitter than the alternatives against which it competed. That is, the process of natural selection that led to the present configuration is one in which genetic differences account for differences in behaviour.

So Alexander sees genetic differences as being crucial to the process of evolution, but environmental differences as characterizing the product of that evolution. He is a type I theorist, since these types pertain to the process of natural selection, not its product.

The distinction between process and product is perhaps a bit harder to grasp when we think of the evolution of some behavioural or psychological trait, but it really applies to any evolutionary event. For the fact of the matter is that evolution driven by a type I selection process feeds on (additive) genetic variation, and uses it up. A morphological character can display the same double aspect. The opposable thumb evolved because there was a genetic difference between those with the thumb and those without it. But once that trait has finished evolving, the difference between those with and those without a thumb may owe more to industrial accidents and harmful drugs taken prenatally than to genetic oddities.

This three-fold division among selection models is of course consistent with there being models that combine two or more of these sorts of process. My taxonomy describes "pure types," so to speak, whereas it is often interesting to consider models in which various pure types are mixed. This is frequently the case in the examples worked out by Cavalli-Sforza and Feldman and by Boyd and Richerson. I want to describe one example from each of these books. The point is to discern the way in which quite different selection processes interact.

In the nineteenth century, Western societies exhibited an interesting demographic change, one that had three stages. First, oscillations in death rates due to epidemics and famines became both less frequent and less extreme. Second, overall mortality steadily declined. This latter change had a multiplicity of causes; improved nutrition, sanitation, and (if the more recent past is also considered) medical advances played a role. The third part of this demographic transition was a dramatic decline in birth rates. Typically, there was a time-lag; birth rates began to decline only after death rates were already on the way down. Cavalli-Sforza and Feldman (p. 181) give the somewhat idealized rendition of this pattern shown in figure 1.

Cavalli-Sforza and Feldman consider the question of how fertility could have declined in Europe. From the point of view of a narrowly Darwinian outlook, this change is puzzling. A characteristic that increases the number of viable and fertile offspring will spread under natural selection, at least when that process is conceptualized from the point of view of a type I model. Cavalli-Sforza and Feldman are not tempted to appeal to the theory of optimal clutch size due to Lack, according to which a parent can sometimes augment the number of offspring surviving to adulthood by having fewer babies.[13] Pre-

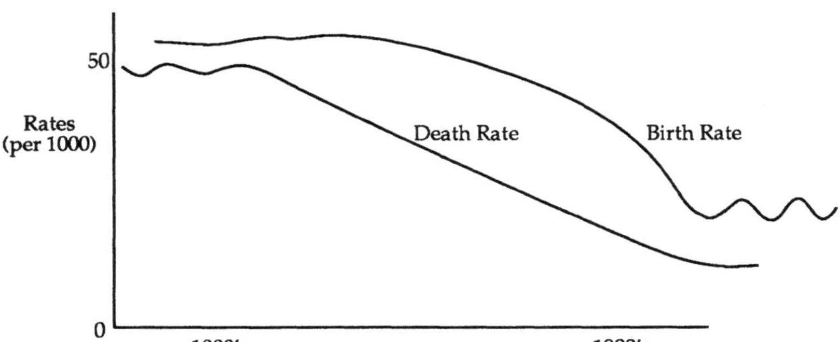

Figure 1. Cavalli-Sforza and Feldman's (p. 181) idealized representation of the demographic transition in Europe. Mortality rates decline; then, after a time lag, the birth rate declines also. (Reprinted by permission of Princeton University Press.)

sumably, this Darwinian option is not even worth exploring, because women in nineteenth century Europe easily could have had more viable fertile offspring than they in fact did. People were not caught in the bind that Lack attributed to his birds.

The trait that increased in the modern demographic transition was one of reduced biological fitness. The trait spread *in spite of* its biological fitness, not *because of* it. In Italy, women changed from having about five children on average to having about two. The trait of having two children, therefore, has a biological fitness of 2/5, when compared with the older trait it displaced.

Cavalli-Sforza and Feldman focus on the problem of explaining how the new custom spread. One possible explanation is that women in all social strata gradually and simultaneously reduced their fertilities. A second possibility is that two dramatically different traits were in play and that the displacement of one by another cascaded from one social class down to the next. The first hypothesis, which posits a gradual spread of innovation, says that fertilities declined from 5 to 4.8 to 4.5 and so on with this process occurring simultaneously across all classes. The second hypothesis says that having five children competed with having two, and that the novel character was well on its way to displacing the more traditional one among educated people before the same process began among less educated people. This second hypothesis is illustrated in Figure 2. There is some statistical evidence that the second pattern is more correct, at least in some parts of Europe.

Cavalli-Sforza and Feldman emphasize that this demographic change could not have taken place if traits were passed down solely from mothers to daughters. The Darwinian disadvantage of reduced fertility is so great that purely vertical transmission is not enough to offset it. This point holds true whether fertility is genetically transmitted or learned. A woman with the new trait will pass it along to fewer offspring than a woman with the old pattern, if a daughter is influenced only by her mother.

What is required for the process is some mixture of horizontal and oblique transmission. That is, a woman's reproductive behaviour must be influenced by her peers and by her mother's contemporaries. However, it will not do for a woman to adopt the behaviour that she finds represented on average in the group that influences her. What is required is that a woman find small family size more attractive than

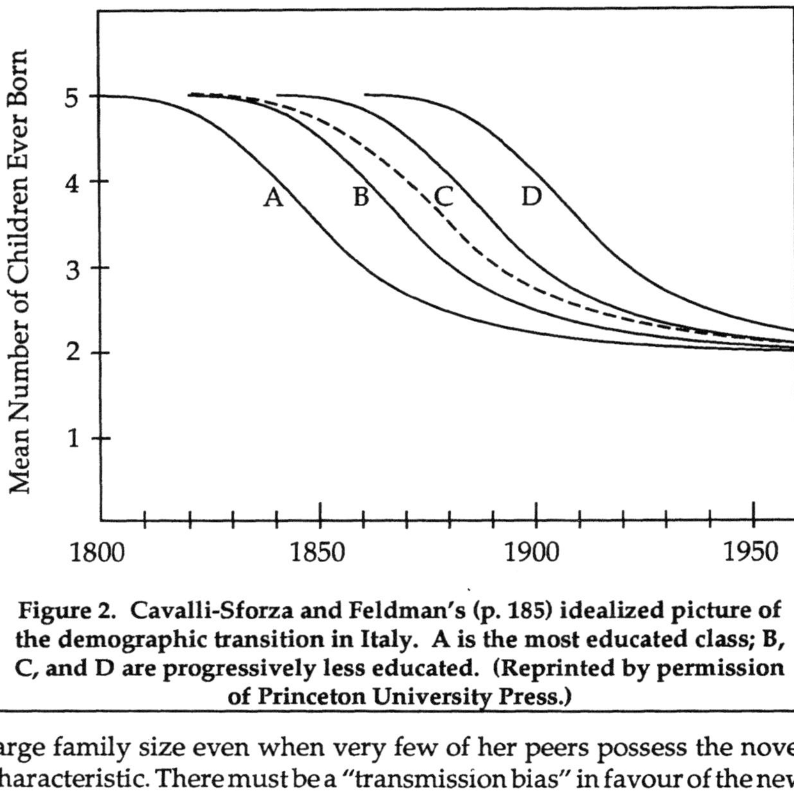

Figure 2. Cavalli-Sforza and Feldman's (p. 185) idealized picture of the demographic transition in Italy. A is the most educated class; B, C, and D are progressively less educated. (Reprinted by permission of Princeton University Press.)

large family size even when very few of her peers possess the novel characteristic. There must be a "transmission bias" in favour of the new trait.

Having a small family was more attractive than having a large one, even though the former trait had a lower Darwinian fitness than the latter. Cavalli-Sforza and Feldman show how the greater attractiveness of small family size can be modelled by using ideas drawn from population genetics. However, when these genetic ideas are transposed into a cultural setting, one is talking about cultural fitness, not biological fitness. So the model they end up with for the demographic transition combines two selection processes. When fitness is defined in terms of having babies, having a small family is selected against. When fitness is defined in terms of the attractiveness of an idea ("having students"), there is selection favouring a reduction in family size. Cavalli-Sforza and Feldman show how the cultural process can overwhelm the biological one; given that the trait is sufficiently attractive (and their models have the virtue of giving this idea quantitative meaning), the trait can evolve in spite of its Darwinian disutility.

The example I want to describe from Boyd and Richerson's book is developed in a chapter that begins with a discussion of Japanese kamikaze pilots during World War II. Self-sacrificial behaviour — altruism — has been an important problem for recent evolutionary theory. Indeed, Wilson called it "the central problem of sociobiology."[14] Although some apparently altruistic behaviours can be unmasked — shown to be predicated on the selfish expectation of reciprocity, for example — Boyd and Richerson are not inclined to say this about the kamikazes. They died for their country. Nor can one explain their self-sacrifice by saying that it was coerced by leaders; kamikaze pilots volunteered. Nor is it arguable that the pilots volunteered in ignorance of the consequences; suicide missions were common knowledge in the Japanese air force.

So why did kamikaze pilots volunteer? Boyd and Richerson (pp. 204-5) refer to one historian who "argues that the complex of beliefs that gave rise to the kamikaze tactic can be traced back to the Samurai military code of feudal Japan which called for heroic self-sacrifice and put death before dishonour. When the Japanese military modernized in the nineteenth century, the officer corps was drawn from the Samurai class. These men brought their values and transmitted them to subsequent generations of officers who in turn inculcated these values in their men."

Boyd and Richerson (pp. 204-5) say that this historical explanation is "unsatisfactory for two reasons. First, it is incomplete. It tells us why a particular generation of Japanese came to believe in heroic self-sacrifice for the common good; it does not tell us how these beliefs came to predominate in the warrior class of feudal Japan. Second, it is not general enough. The beliefs that led the kamikazes to die for their country are just an especially stark example of a much more general tendency of humans to behave altruistically toward members of various groups of which they are members." They then impose two conditions of adequacy on any proposed explanation: (i) it must show

how the "tendency to acquire self-sacrificial beliefs and values could have evolved;" (ii) it must show "why altruistic cooperation is directed toward some individuals and not others (p. 205)."

In answer to these requirements, Boyd and Richerson then construct a group selection model that incorporates a certain form of learning. Altruists and selfish individuals exist in each of several groups. Within each group, altruists do less well than selfish people. However, groups of altruists go extinct less often and found more colonies than groups of selfish individuals. These ideas are standard fare in the models of group selection that evolutionary biologists have considered.[15] A type I selection model of the evolution of altruism will require a between-group process favouring altruism that offsets the within-group process that acts to eliminate the trait.

The new wrinkle introduced by the idea of cultural transmission is as follows. Boyd and Richerson postulate that cultural transmission favours common characteristics and works against rare ones. Within a group, individuals are especially biased towards adopting altruism if most individuals are altruists and towards becoming selfish if most people are selfish. What I mean by "especially" biased is illustrated in Figure 3. In all cases of cultural transmission, the state that a naive individual acquires is influenced by the frequency of traits in the population. Boyd and Richerson impose a more extreme demand. They require that the probability of acquiring a common trait be higher than its population frequency; this is what they call "frequency-dependent biased transmission" (depicted in Figure 3c).

The process of cultural transmission can work within the timeframe of a single biological generation. The effect is to augment the amount of variation there is among groups. Whereas traditional genetic models of group selection allow for a continuum of local frequencies of altruism, the result of this biased transmission rule is to push each local population towards 100% altruism or 100% selfishness. This has the effect of raising the probability that altruism will evolve and be maintained.

Boyd and Richerson also raise the question of how this biased "conformist" transmission rule could have evolved in the first place. They speculate that if a species is composed of a set of local populations, and if these populations inhabit qualitatively different micro

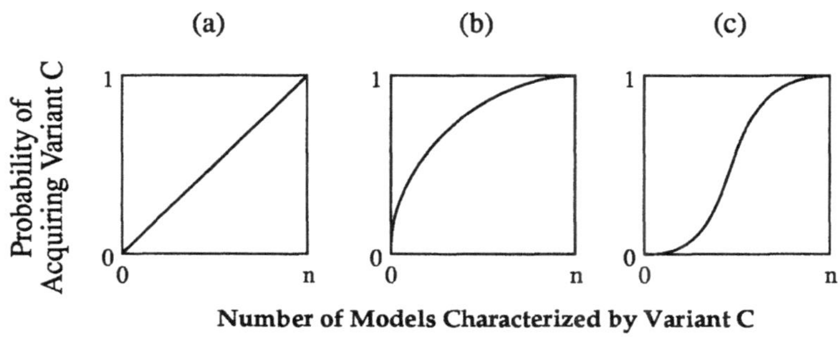

Figure 3. Boyd and Richerson's (p. 207) characterization of three patterns of cultural transmission. In all three cases, the probability that a naive individual will acquire a trait depends on the frequency of the trait among the individual's models. (a) represents unbiased transmission, (b) directly biased transmission, and (c) frequency-dependent biased transmission. (Reprinted by permission of University of Chicago Press.)

habitats, an individual moving into a new habitat may do best by imitating the traits that are common there. Their proposal is a Darwinian explanation for acting Roman in Rome, so to speak. Once this transmission bias has evolved, it may have various spin-off consequences that have the effect of harming organisms rather than helping them. If you find yourself living with altruists, the transmission bias will lead you to become altruistic yourself, even though you would be better off remaining selfish. Boyd and Richerson admit that there is little or no psychological evidence that people deploy the extreme form of transmission bias that their model postulates.

Just as in the example discussed from Cavalli-Sforza and Feldman, this model of Boyd and Richerson's mixes together the concepts of biological and cultural fitness. Altruism is deleterious to individuals, when fitness is calibrated in terms of the survival and reproduction of organisms. But common characteristics are more contagious than rare ones, when the individuals use a conformist transmission rule. This means that when altruism is common, it is more catching than selfishness. In such cases, the cultural fitness of altruism is greater than the cultural fitness of selfishness, when one considers a group in which altruism is common. The net result is that the special cultural transmission rule can allow a characteristic to evolve that could not evolve without it. Within purely biological models, altruism is eliminated in a large range of parameter values. The prospects for altruism to evolve are enhanced when culture is included in the model. Just as in Cavalli-Sforza and Feldman's discussion of the demographic transition, assumptions about cultural transmission lead to predictions that would not be true if a purely biological and noncultural process were postulated.

The two examples I have described are typical of the models discussed in the two books. The models aim to show how different patterns of cultural transmission make a difference for how a psychological or social characteristic will evolve. Although most of the emphasis is placed on identifying cultural analogs of natural selection, the authors do develop nonselective models of cultural change. For example, population geneticists have described how genes of nearly identical fitness can change frequency in a population by doing a random walk. The models developed for random genetic drift, as it is called, can be used to describe the process by which family names disappear. This helps explain why the descendants of the Bounty mutineers have come to share the same surname. A reduction in variation is the expected consequence of random walks, both genetic and cultural.[16]

What are we to make of the research program embodied in these books? Biologists interested in culture are often struck by the absence of viable general theories in the social sciences. All of biology is united by the theory of biological evolution. Perhaps progress in the social sciences is impeded because there is no general theory of cultural evolution. The analogies between cultural and genetic change are palpable. And at least some of the disanalogies can be taken into account when the biological models are transposed. For example, the Weismann doctrine tells us that variation is "undirected;" mutations do not occur because they would be beneficial. But ideas are not invented at random. Individuals often create new ideas — in science, for example — precisely because they would be useful.[17] Another and related disanalogy concerns the genotype/phenotype distinction. An organism's genotype is a cause of the phenotype it develops; that same genotype also causally contributes to the genotype of the organism's offspring. But there is no further pathway by which a parental phenotype can causally shape the genotype of its offspring. This is one way of describing the idea that there is no "inheritance of acquired characteristics." No such constraints seems to apply to the learning that occurs in cultural transmission.

These disanalogies between genetic and cultural change do not show that it is pointless or impossible to write models of cultural evolution that draw on the mathematical resources provided by evolutionary theory. In a sense, it is precisely because of such differences that there is a point to seeing the consequences of models that take these differences into account. These structural differences between genetic and cultural evolution do not undermine the idea that models of cultural evolution have a point.

Another reservation that has been voiced about models of cultural evolution is that they atomize cultural characteristics. Having two children rather than five, or being a kamikaze pilot, are characteristics that are abstracted from a rich and interconnected network of traits. The worry is that by singling out these traits for treatment, we are losing sight of the context that gives them cultural meaning.

It is worth mentioned that precisely the same question has been raised about various models in genetic evolution itself. If you wish to understand the population frequency of sickle cell anaemia, for example, you cannot ignore the fact that the trait is correlated with resistance to malaria. In both cultural and genetic evolution it is a mistake to think that each trait evolved independently of all the others. Of course, the lesson to be drawn from this is not that one should not atomize characteristics, but rather that the atoms one identifies should be understood in terms of their relationship to other atoms.

In fact, this emphasis on context is one of the virtues that Boyd and Richerson think their approach has over the approach taken by sociobiology. According to the models under review, genetic selection has given our species the ability to engage in social learning. Once in place, this cultural transmission system then allows characteristics to evolve that could not have evolved without it. In other words, it is only because the traits in question evolve in the context of a cultural transmission system that they are able to evolve at all.

We need to recognize that the descriptors singled out for treatment in science always abstract from complexities. If there is an objection to the descriptors used in models of cultural evolution, it must concern the details of how these models are constructed, not the mere fact that they impose a descriptive framework of some sort or other.[18]

Although the criticisms I have reviewed so far do not seem very powerful, there is a rather simple fact about these models that does suggest that they may be of limited utility in the social sciences. In so far as these models describe culture, they describe systems of cultural transmission and the evolutionary consequences of such systems. Given that the idea of having two children was more attractive than the idea of having five, and given the horizontal and oblique transmission systems thought to be in place, we can see why the demographic transition took place. But as Cavalli-Sforza and Feldman recognize, their model does not begin to describe why educated women in nineteenth century Italy came to prefer having smaller families, or why patterns adopted in higher classes cascaded down to lower ones. The model describes the consequences of an idea's being attractive, not the causes of its being attractive.

This distinction between the consequences of fitness differences and the causes of fitness differences also applies to theories of biological evolution.[19] A population geneticist can tell you what the evolutionary consequences for a population will be, if the genes in the population bear various fitness relationships to each other. It is a separate question to say why a given gene in fact is fitter than the alternatives. For example, consider the simplest of one-locus two-allele models for a diploid population. There are three genotypes possible at the locus in question, which we might label AA, Aa, and aa. If the heterozygote genotype is fitter than the two homozygote forms, the population will evolve to a stable polymorphism. Neither allele will be eliminated by the selection process. This is a simple algebraic fact, one having nothing to do with the biological details of any living population. Models such as this one can be thought of as intellectual resources that biologists interested in some particular population might find reason to use.

When human geneticists apply this model to the sickle cell system, they say that Aa is the fittest genotype because heterozygotes at the locus in question have enhanced resistance to malaria and little or no anaemia. The two homozygotes have lesser fitnesses because they are either anaemic or lack the malaria resistance. These specific remarks about the locus in the relevant human population describe the sources of fitness differences. Alternatively, a fruitfly geneticist may take the same population genetics model and apply it to a locus in some Drosophila population by saying that the heterozygote has enhanced temperature tolerance over the two homozygotes. The population consequences of heterozygote superiority are the same in the two cases; a stable polymorphism evolves. It is the sources of the fitness differences that distinguish the human application from the application to fruitflies.

This, I think, is the main shortcoming of the models of cultural evolution I am considering. The illumination they offer of culture concerns the consequences of cultural transmission systems. But there is far more to culture than the consequences of the rules that describe who learns what from whom. Social scientists have not wholly ignored the way that patterns of influence are structured in specific cases. An historian of nineteenth century Italy might attempt to explain why some traits found among educated people were transmitted to lower social strata, while others were not. Again, it is the sources of the transmission system that will interest the social scientist. The social scientist will take it for granted that the consequences of this influence will be that ideas cascade from one class to another.

Models of transmission systems describe the quantitative consequences of systems of cultural influence. Social scientists inevitably make qualitative assumptions about the consequences of these systems. If it could be shown that these qualitative assumptions were wrong in important cases, and that these mistakes actually undermine the plausibility of various historical explanations, that would be a reason for social scientists to take a greater interest in these models of cultural evolution. But if the qualitative assumptions turn out to be correct, it is perhaps understandable that historians should not accord much importance to these investigations.[20]

Population genetics really is a unifying framework within evolutionary theory. Fruitflies and human beings differ in many ways, but if a one-locus system exhibits heterozygote superiority, the population consequences will be the same, regardless of whether we are talking about people or Drosophila. Evolutionary theory is much less unified when we consider what it has to say about the sources of fitness differences. There are many, many models that treat a multiplicity of life-history characteristics and ecological relationships. Evolutionary theory achieves its greatest generality when it ignores sources and focuses on consequences.

The transposition of evolutionary models to the social sciences is a transposition of the most unified and complete part of evolutionary theory, one that leaves behind less unified theoretical ideas. This is not a criticism of the models of cultural evolution that result, but a fact about the price one pays for very general theorizing of this type. Cultural learning is a cultural universal. And patterns of cultural learning conveniently divide into vertical, horizontal, and oblique subcases. When ideas differ in their attractiveness, the system of transmission will determine the rate of change and the end- state that the population achieves. Only because they develop theories *within* this narrow compass do these models *of* cultural evolution have the generality they do.

Many of the examples discussed in the two books I have been considering describe evolution within a culture, not the evolution of the cultural transmission system itself. However, Boyd and Richerson, especially, also concern themselves with the way a system of cultural learning could have evolved by straightforward Darwinian means. Here the authors are not giving a model of how human cultures work, once they exist, but are trying to show how cultural learning became a possibility in the first place. This project obviously is a very important one, but not one that applies to many social scientific research programs. A correct genetic explanation of this important feature of the human phenotype would not provide a unifying framework within which social scientists would then do their work. They would not use this theory at all. It is one thing to explain the demographic transition in nineteenth century Italy, something else to explain why human beings are able to learn from individuals who are not their biological parents.

In spite of these shortcomings, there is a basic achievement of these models of cultural evolution that deserves emphasis. A persistent theme in debates about sociobiology, about the nature/nurture controversy, and in other contexts as well is the relative "importance" that should be accorded to biology and culture. I place the term "importance" in quotation marks to indicate that it is a vague idea crying out for explication. Nonetheless it has been a fundamental problem in these controversies to assess the relative "strength" or "power" of biological and cultural influences.

One virtue of these models of cultural evolution is that they place culture and biology into a common framework, so that the relative contributions to an outcome are rendered commensurable. What becomes clear in these models is that in assessing their relative importance of biology and culture, *time is of the essence.* Culture is often a more powerful determiner of change than biological evolution because cultural changes occur faster. When biological fitness is calibrated in terms of having babies, its basic temporal unit is the span of a human generation. Think how many replication events can occur in that temporal interval when the reproducing entities are ideas that jump from head to head. Ideas spread so fast that they can swamp the slower (and hence weaker) impact of biological natural selection.

There is a vague idea about the relation of biology and culture that these models help lay to rest. This is the idea that biology is "deeper" than the social sciences, not just in the sense that it has developed farther, but in the sense that it investigates more fundamental causes. A social scientist will explain incest avoidance by describing the spread of a custom; the evolutionary biologist goes deeper by showing us why the behaviour evolved. The mind set expressed here is predisposed to think that culture is always a weak influence when it opposes biology. The works described here deserve credit for showing why this common opinion rests on a confusion.

In spite of this achievement, I doubt that these models of cultural evolution provide a general framework within which social scientific investigations may proceed. My main reason for scepticism is that these models concern themselves with the consequences of transmission systems and fitness differences, not with their sources. Social scientists interested in cultural change generally focus on sources and make do with intuitive and qualitative assessments of what the consequences will be. It isn't that the biologists and the social scientists are in conflict; rather, they are talking past each other.

Dobzhansky is famous for having said that "nothing in biology can be understood except in the light of evolution." His idea was not the modest one that evolution is necessary for full understanding; that would be true even if evolution's contribution were minor, though ineliminable. Rather, Dobzhansky had in mind the stronger claim that evolutionary considerations should be assigned pride of place in our understanding of the living world. A transposition of Dobzhansky's slogan to the topic of this paper would say that "nothing in the social sciences can be understood except in the light of models of cultural evolution." My suspicion is that only the weaker reading of this pronouncement is defensible.

NOTES

[1] I worked on this paper while a William Evans Fellow at the University of Otago during parts of July and August, 1990; my thanks to the University and to the members of the Philosophy Department for inviting me and for making my stay such an enjoyable one. This paper expands upon a talk I gave in December, 1985 at the University of Palma de Mallorca called "Natural Selection and the Social Sciences." I'm grateful to Robert Boyd, Dan Hausman, Peter Richerson, and David S. Wilson for comments on an earlier draft.

[2] John Maynard Smith (1982), *Evolution and the Theory of Games*, Cambridge University Press.

[3] L. Cavalli-Sforza and M. Feldman (1981), *Cultural Transmission and Evolution: A Quantitative Approach*, Princeton University Press.

[4] R. Boyd and P. Richerson (1985), *Culture and the Evolutionary Process*, University of Chicago Press.

[5] "Having babies" should be interpreted broadly, so as to include "having grandbabies", "having greatgrandbabies", etc. In some selection models (e.g., Fisher's sex ratio argument), fitness differences require that one consider expected numbers of descendants beyond the first generation.

[6] See R. Colwell and M. King (1983), "Disentangling Genetic and Cultural Influences on Human Behaviour: Problems and Prospects," in D. Rajecki (ed.), *Comparing Behavior: Studying Man Studying Animals*, Lawrence Erlbaum Publishers.

[7] These are reviewed in J. Hirshliefer (1977), "Economics from a Biological Viewpoint," *Journal of Law and Economics* **1**: 1-52.

[8] See K. Popper (1973), *Objective Knowledge*, Oxford University Press.

[9] A variety of "selective-retention" models of learning and of scientific change are reviewed in Donald Campbell (1974), "Evolutionary Epistemology," in P. Schilpp (ed.), *The Philosophy of Karl Popper*, Open Court Publishing. David Hull's *Science as a Process* (University of Chicago Press, 1988) develops some interesting ideas about how evolutionary ideas can be used to explain scientific change.

[10] I do not claim that this taxonomy is exhaustive. For example, the spread of an infectious disease may be thought of as a selection process, in which the two states of an individual ("infected" and "not infected") differ in how catching they are. Clearly, this is not a type I process. Arguably, the concept of learning does not permit this process to be placed in type II. Perhaps the taxonomy would be exhaustive, if "learning" were replaced by "phenotypic resemblance not mediated by genetic resemblance."

[11] The description of type III models, in which fitness is measured by "having students," is due to Peter Richerson.

[12] See, for example, Richard Alexander (1979), *Darwinism and Human Affairs*, University of Washington Press.

[13] See D. Lack (1954), *The Optimal Regulation of Animal Numbers*, Oxford University Press.

[14] E. Wilson (1975), *Sociobiology: The New Synthesis*, Harvard University Press.

[15] See E. Sober (1988), "What is Evolutionary Altruism?" New Essays on Philosophy and Biology (*Canadian Journal of Philosophy Supplementary* Volume **14**), University of Calgary Press.

[16] See L. Cavalli-Sforza and M. Feldman, Ibid., pp. 255-66.

[17] The difference between directed and undirected variation is conceptually different from the difference between biased and unbiased

transmission. The former concerns the probability that a mutation will arise; the latter has to do with whether it will be passed along.

Directed variation (mutation) can be described as follows. Let u be the probability of mutating from A to a and v be the probability of mutating from a to A. Mutation is directed if (i) $u>v$ and (ii) $u>v$ because $w(a) > w(A)$, where $w(X)$ is the fitness of X.

[18] See J. M. Smith (1989), *Did Darwin Get it Right?*, Chapman and Hall.

[19] See Elliott Sober (1984), *The Nature of Selection*, MIT Press.

[20] So the question about the usefulness of these models of cultural evolution to the day-to-day research of social scientists comes to this: Are social scientists good at intuitive population thinking? If they are, then their explanations will not be undermined by precise models of cultural evolution. If they are not, then social scientists should correct their explanations (and the intuitions on which they rely) by studying these models.

Elliott Sober
Philosophy Department
University of Wisconsin

Kim Sterelny

PUNCTUATED EQUILIBRIUM AND MACROEVOLUTION

1. INTRODUCTION

Hardened Darwinism is the view that the history of life is explained by forces operating on populations of organisms, or perhaps, on their genes. Large scale events and processes are no more than an aggregation of the fate of individual organisms. This story has received some rough treatment in the literature lately, for hierarchical views of evolutionary theory are currently popular. Eldredge, Gould and Stanley[1], to name only three, are on the record in urging that there are robust macroevolutionary phenomena, phenomena not captured generation by generation change in gene frequency. Three large scale patterns are most on people's minds. One is mass extinction. Mass extinctions, its been alleged, are regular and intense. Most importantly, surviving mass extinction is mostly chance; it is not determined by an organism's suite of adaptations. If so, the shape of the tree of life is determined not by the relative fitness of its various twigs, but by their proximity to an extrinsic, perhaps even extraterrestrial, pruner. A second is Gould's most recent preoccupation, the "cone of decreasing diversity". The theme of his 1989 is his idea that life reached its maximal diversity at the Cambrian explosion; diversity has declined since. On Gould's view, natural selection played no central role in either the establishment of that diversity or its decline[2]. But my focus is the hypothesis of "punctuated equilibrium". At a minimum, this is the claim that the fossil record reveals that a species' typical life history is rapid formation followed by stasis for the bulk of its life span, then disappearance by extinction or by splitting into daughter species.

P. Griffiths (ed.), Trees of Life, 41–63.

Species do not typically form by gradual transformation out of their ancestors and disappear by gradual transformation into their descendants. This pattern is alleged to have two striking consequences. In some of their early work, Gould[3], Eldredge and Stanley have taken it to have implications about the process of change. The changes we see in the disruptions of stasis are too rapid and too large to be effected by a generation by generation shift of gene frequencies in a population. The changes are not effected by natural selection working on normal variation in a population. Species are established by something like macromutation. Moreover, the stability of species require explanation. Gould has written of the "paradox of the first tier" (1983, 1985). If natural selection were the only force operating on populations, we would expect them to converge on optimum design. Since they do not, there are higher level forces in action. More recently, Eldredge in particular (but also Gould) have seen the hypothesis of punctuated equilibrium as part of a theory of biological hierarchy. If the pattern hypothesis is right, species are subjects in their own right of evolutionary laws and causal explanations. The role and fate of a species is not a mere summation of the role and fate of its members. Let me try to unpack this a little, concentrating to begin with on the different ways the punctuational hypothesis can be taken.

2. PUNCTUATED EQUILIBRIUM: THREE VERSIONS

2.1. PATTERN HYPOTHESES

We can take the punctuational hypothesis as just the claim that the typical life cycle of a species is one of quick origin, morphological stability, and rapid disappearance. So understood, its two crucial empirical claims are that the apparent gappyness of the fossil record is real, no mere artifact of a very incomplete record, and that the tempo and mode of evolution is not uniform.

If the pattern is genuine, it poses some difficult questions: (a) how is stasis maintained (b) how and why does it break down (c) what are the mechanisms of the punctuations (d) what then is the nature of species and speciation? In answering these questions, the idea runs, we must move beyond the idea that evolution is nothing but the gradual change in the composition of a population, and the associated fictionalist view that species names are of arbitrary segments of a lineage that changes fairly smoothly over time.

There is doubt about the evidence for the pattern. There is good reason to expect stasis to be over-estimated. Species with a broad range and/or a large population are more likely to be exemplified in the fossil record. But they are also more likely to exhibit stability in form and behaviour. Moreover, there is a hard part bias. Almost always, only hard parts are fossilized. The transformations over time of soft parts, biochemistry and behaviour largely escapes the record. So fossil traces will preserve the variation of only a fragment of a species traits. Moreover, since so few organisms leave fossils, the record of a species is like to understate variation in that species.

Despite the problems of vindicating the pattern hypothesis, I am going to assume its true, and will be explore its ramifications.

2.2. PROCESS HYPOTHESES[4]

Considerations about punctuated equilibria have been supposed to fracture hardened Darwinism in two ways. First, it has been supposed that change requires a saltation, a change from one species to an other without intermediate forms[5]. Second, it has been supposed that stasis, the stability of a species' form over long periods requires explanation; an explanation the neodarwinian cannot give, since that theory predicts not stasis but slow change.

2.2.1. Change

There is a standard argument against macromutations. They are possible, but are almost certain to be unfavourable. For most points in genetic space are unviable. Viable combinations are islands in a huge see of unviability. Any organism is on the shores of one of these islands, though perhaps not on quite the highest point. A small change might take it up the slope a bit. But a large change, unless by very rare chance it finds a new island, will doom the organism's offspring.[6] Moreover, Mayr has an alternative theory of fast change. A geographically and reproductively isolated population at the periphery of its normal range is likely to evolve quickly and without fossil trace. For such groups are small, so drift and natural selection can work more quickly. Moreover, these isolates are under strong selection pressure, for they are not in their normal niche, hence are not well adapted. Since they are isolated, changes are concentrated, and are not diluted by migration[7].

Evolutionary change can be rapid in those circumstances, yet the intermediate forms are well adapted to neither their former niche nor the one now being colonized. So they will be neither long lasting nor widely distributed. So its no surprise that these intermediaries are not found in the fossil record. This hardened story is so convincing that these days Gould denies that he ever had macromutation in mind as an account of punctuational change, a denial that sits uneasily with a good deal of his early rhetoric and of his many complimentary references to Goldschmidt.

2.2.2. Stability

Gould has argued that the stability of species constitutes a threat to hardened Darwinism. He thinks that morphological and developmental constraints sharply limit the power of natural selection to change a species' basic structural design. These constraints can break down only in special circumstances. Gould conceives of the hardened Dominant as implicitly thinking of natural selection as a mighty transforming agent, slowly but inexorably shaping an organism to its niche. That view of selection suffers from a "paradox of the first tier" (Gould 1983, 1985). If natural selection were the most important force shaping the biosphere, we should see life being pushed in the direction of greater optimality. But we do not. Species usually do not change. Moreover, there is no sense in which life is better adapted now that in (say) the time of the dinosaurs. So stasis is an anomaly for those who expect natural selection to optimize organic design. So Gould urges us to see punctuated equilibrium as not just a claim about pattern, but one about process as well. Evolution involves the differential survival of species, not just their members.

I don't buy any of this. I am not convinced that stasis is an an anomaly for those who think that natural selection is the only selective force is operating on on individuals in a population. Moreover, neither species nor any other high level selection is a suitable explanation of stasis. Let me take these ideas in turn.

There are good reasons for expecting species to be stable. Stability can be the result of habitat tacking. As the environment changes, organisms may react by tracking their old habitat. They might move north as the climate cools, rather than by evolving adaptations to the cold. Selection will usually drive tracking. For migrants that follow the habitat (personally or by reproductive dispersion) will typically be fitter than the population fragment that fails to move, for the residual fragment will be less well adapted to the new environment and will be faced with new competition from other migrants tracking their old habitat. The evolution of developmental canalization can stabilize the phenotype, as does selection itself. As a population approaches optimum fitness for the particular fitness island on which it finds itself, deviations from that optimum are punished, even if the island is only a local optimum, and even if there is a superior genotype quite close to the local best.

Moreover, we need not invoke high level selection to explain imperfect design. We know natural selection is not the only force operating on the genetic make-up of a population; drift is another. Moreover, if the day by day processes on populations are calamitously disrupted every 28 million years or so, whatever optimizing processes there will be disrupted too, but not because of a higher level selective force.

Furthermore, we should distinguish adaptation from optimization. I can think that every trait is an adaptation without thinking that any are optimal. Thick fur, for example, is an adaptation for insulation if creatures with thick fur have survived longer or reproduced more fecundly in virtue of its insulating properties than some of their rivals. Nothing in this story presupposes optimality; some of the competition may have succumbed by ill luck; drift and selection might combine to explain the widespread possession of fur. Moreover, the fur is an adaptation for insulation even if it is not an optimal adaptation to insulation; others may have developed both fur and fatty insulation under the skin[8].

So even if natural selection is the only selection, and even supposing it to act everywhere and always, we would not expect organisms to have, or to be approaching, perfect design. Moreover, even if life were locally optimal, we would not expect the history of life to be a history of progress. For optimal adaptation is relative both to local conditions and the constraints imposed by the species' history and engineering. Organisms in radically differing environments, or with radically differing designs, cannot be compared for goodness of design. On both counts, it makes no sense to say that (eg) sharks are better adapted than tigers. Yet the picture of life's history as a progress requires just this empty decontextualized notion of optimality. Not even a very optimistic adaptationist, therefore, predicts a long succession of ever better adapted biota. They do not expect a march of progress, so there is no need to posit an extra, high level, mechanism to disrupt the march.

Stasis may well be real; no artifact of the record, but still be a by-product of common or garden natural selection.

But imagine we do need to explain why microevolutionary processes fail to transform a species. The shift from thinking of punctuation as a "Goldschmittian" process to the picture of it as providing the variation for a high level selective force robs the defender of punctuated equilibrium of the resources to explain stasis. If punctuation is an event or process that somehow has the powers to crack the shell of homeostatic processes, architectural and developmental constraints that normally grip a species, then it is a candidate explanation of stasis. We explain it by explaining why those events are rare. But this is to see stasis and its breakdown as a variant process at the level of organisms in a population. Species selection cannot explain stasis since stasis and its breakdown will be part of the theory of variation on which species selection acts. Just as the rate of mutation is part of the boundary conditions of natural selection, rather than something natural selection explains (except perhaps in the very long term), stasis and its breakdown will be part of the boundary conditions of species selection. In any case, adding an extra selective force will not explain why things stay the same unless miraculously, the extra force automatically cancels out the first.

So just as we don't need any fancy mechanism to explain change, we don't need to depart from hardened Darwinism to explain stability either.

2.3. SPECIES AS AGENTS

Macroevolution does not stand or fall with fancy mechanisms. An analogy from the philosophy of psychology might make this clear. Cartesian dualists, no doubt, think that psychological processes are irreducible to neurophysiological ones. But functionalists have shown you do not have to be a dualist to think that psychology has a limited but significant autonomy. Each psychological mechanism is neurally realized; we must be able to give a neural explanation of how the mechanism works. But, the idea runs, we would lose explanatory power if we dispensed with psychological explanation in favour of neural ones. Why this is so is a matter of considerable debate. One popular idea is that a psychological mechanism can have distinct neural realizations. So dispensing with psychology would dissolve a unitary phenomenon into myriads of subcases. However this debate goes, the psychology-neuroscience relation gives us a rough model of how macroevolutionary processes might be irreducible to changes in populations of organisms, even though they are realized by, and composed from such changes. It gives us a rough model of how macroevolutionary explanations might be indispensable yet not require bizarre mechanisms.

I think we can extract a reasonable macroevolutionary thesis in which the punctuational hypothesis plays a central role[9]. For it gives aid and comfort to the view that species are agents in the evolutionary story, by (i) showing that species can be identified and by (ii) showing them to have a causal role. Consider first identification. If the pattern hypothesis is true, species' names do not merely pick out arbitrary segments of a continuum of variation; they do not come into existence by transformation out of their ancestor. Even though the process is not literally instantaneous, species speciate. They come into existence through splitting. It does not even matter if they change between formation and disappearance so long as those changes do not constitute speciation. Species on this view are identifiable segments of a lineage.

Moreover, species are important because speciation is important. Morphological transformation, the invention and establishment of new adaptations occurs at the same time, and may even be the same process, as the division of a species into daughter species. Furthermore, if lineages do not change by transformation, then long term trends in lineages can hardly be the result of its slow transformation. But there are many examples of such trends. Famous examples are size increase in the horse lineage, and brain size increase in the homo lineage. So perhaps these trends are caused by differential extinctions of species in the lineage. Horse species of various sizes have been born, but the larger species have survived better.

3. SPECIES SELECTION

We can see the theory of punctuated equilibria as a plausible revision of orthodoxy if we see it as showing that species are agents. It would thus be an argument for seeing a hierarchical structure in evolutionary history. One way of emphasizing the significance of species is to argue that evolutionary history is in part the story of the differential selection of species, not just their members. But we must at least distinguish between species sorting and species selection[10]. Species sorting is the mere differential survival of species with some particular characteristic. No claim about the causal import of that characteristic need be made. Everyone agrees that species are sorted; that, for example, species with the distinctive features of trilobites have failed to survive. No one, as Philip Kitcher has reminded me, has ever denied the existence of macroevolutionary effects; the radical suggestion is that there are macroevolutionary causes. Species selection is this much more robust claim; a species' fate is affected by some property of the species itself, not just the properties of the organisms that compose it. But even this characterization of the issues is too crude.

David Hull and Richard Dawkins have both pointed out that talking of genes, organisms or species as selective agents is ambiguous[11]. There are two components to evolution under selection: replication and interaction. For there to be evolution there must be sequences of replicators, of entities that are copied and, through being copied, pass on information-bearing structure to the next member of the sequence. Successful replicators in this way root long and bushy lineages. On both Hull's and Dawkins' view, evolution is the differential growth of lineages. But lineages grow, or fail to, in part in virtue of their interaction with the environment. The elements from which they are formed need to maintain their structural integrity, gather resources, and replicate through interaction with their environment. In some evolutionary regimes, the two components of the evolutionary process are matched by a specialization of entity to role. There is not just replication and interaction, but also replicators and interactors, or vehicles, though this specialization of function cannot be complete; replicators must interact, though interactors need not replicate. So Dawkins, for example, argues that though organisms are vehicles, and in that sense are units of selection, they do not replicate.

So, in confronting the suggestion that species are units of selection, we need to consider whether they are units of replication or interaction or whether, without specialization, they do both. Orthodoxy is clearly under threat if species are interactors; it may be stretched even if species are active replicators.

4. DO SPECIES REPLICATE?

I do not think it should be very controversial that we can see species as replicators. It is much less obvious that we must so see them. The significant structure passed on in speciation is not range or habitat, but the information in the species gene pool[12]. It is true that an isolated fragment of a population undergoing speciation will be less genetically diverse than the parent pool, and will be in some way modified by selection. So species are rather crude replicators, passing information across a speciation with some but not drastic modification. It does not follow that from this alone that we need to see species as replicators. Some of the actual and possible examples of the recent literature are, I think, best read as exploring the idea that it is essential to see species as replicators. Maynard Smith and Sober[13], for example, give potential examples of "species selection" in which the causally relevant property is a trait of the individuals in the species. Altruistic behaviour is often considered a candidate for explanation by high level selection; so too is the prevalence of sexual reproduction. Sober considers the effect of individual dispersal capacities on the propensity to speciate. In all these cases, the traits in question are those of individual organisms. The diagnostic question, in filtering out replicator selection, is "Who benefits?". In Maynard Smith's and Sober's examples, the benefit falls to a species, and we benefit in seeing a species as a replicator, a replicator whose replication potential is increased if the individual organisms composing it have certain characteristics. What is to be explained is the surprising prevalence of species composed of organisms that reproduce sexually, or the existence of species whose members behave altruistically. Seeing species as replicators is mandatory, it has real explanatory bite, if the prevalence of sex or existence of altruism has no other explanation. We are then required to expand our picture of the evolutionary process: not just genes and genomes, but also species are amongst the undoubted replicators. But if Kitcher and I were right in our 1988, there can be alternative good pictures of a selective process. So one might see species as replicators even if sex, altruism, or dispersal characteristics can be otherwise explained. For that reason the idea that species are replicators just in itself may not revise the orthodox tale of evolution.

5. DO SPECIES INTERACT?

We move beyond the synthesis if we see species not just as replicators but also as vehicles. For then they are not just the scoreboard but players in the game. If the diagnostic question in detecting replicator selection is "Who benefits?", then in detecting vehicle selection it is "Who acts?". If they act, who do they act for? We can safely assume that their speed of reaction will be slow; they can scarcely be seen as vehicles of particular organisms. If they are interactors, the replicator they benefit is the species as a whole, or some substantial fragment of it. So if species are vehicles, there must as well be high level replicators. For a species to act, three requirements must be met. (i) They must be biologically coherent entities. They must be located in time, space and ecological context. They must have some internal organization. (ii) A species must stand to its gene pool in something like the way an organism stands to its genome. Species must have something like a phenotype: it must have characteristics through which it interacts with the world in order to assist its own copying or that of some other replicator. (iii) We need an account of the empirical and conceptual distinction between, for example, a species going extinct because of species selection in favour of large-brain hominids, and a species going extinct because its small brain members are outcompeted by the members of a larger brained species. To draw this distinction, there must be phenomena explicable only by appeal to species-level causal processes.

6. ARE SPECIES COHERENT?

If species are vehicles they must be biologically coherent[14]. They must have temporal and physical bounds. Extinction imposes a reasonably determinate end. If speciation is splitting from an ancestral species then species have reasonably distinct beginnings too. Eldredge points out that species have distribution limits which impose a spatial bound analogous to an organism's skin. It is less clear that they are ecologically localised. It is common to speak of species being adapted to particular ecological niches, but it is typically false. The communities into which a species is divided often have quite different life habits and find their way into different environments. Think, eg, of differing populations of deep bush and suburban adapted possums in Australia, or the differences between Australian and feral New Zealand possums. If a species of possum has found its way into differing niches, there may be no selection force operating on the possum species as such. This fact is not fatal to the idea of high level vehicles. Selection for low population density might preserve one species of desert dwelling numbat, and drive a more densely congregating one to extinction, even though not all numbats of that surviving species were desert dwellers. So Eldredge now thinks of avatars, a breeding fragment of a species in a niche, as his candidate for a supra-individual unit of selection. By this he had better mean that avatars are vehicles acting to enhance the replication of the species from which they are drawn, for avatars are even cruder replicators than species. I have mentioned that species are only replicators in a rather rough sense. The gene pool of the daughter species is distinct from that of the parent through both drift and through selected changes in genetic composition. Avatars are replicators in still rougher a sense, for the next generation is unlike the current one not just through selection and drift but also immigration. Avatars, unlike speciating fragments, are not ex officio isolated from other populations. So avatars, perhaps unlike species, are physically, temporally and ecologically localised enough to be agents.

But have avatars or species any structure, any internal organization? Noone would pretend that they have the complex internal organization characteristic of complex multi-celled animals, but Hull, for example, warns us against this zooaphilia. Metazoans are far from the typical organism. Still, if species (or avatars) act, there must be some glue that holds them together, something in virtue of which the organisms that make up a species are part of a whole. Eldredge, in his 1985a, argues that the "reproductive plexus" binds members of a species together. But this idea faces many problems. If we rely on actual genealogical history, individuals near a speciation event will be closer in descent line to organisms that are not species mates than to some conspecifics. It is difficult to apply the intuitive notion of a species as a reproductive community[15]. Some individuals that could interbreed do not. We have special problems applying the notion of a reproductive community to geographically scattered species divided up into races, to temporally scattered fragments of a species, and to species whose members have strong mating preferences. Only the first two of these difficulties are eased by refocusing on avatars rather than species. The notion of a reproductive plexus is in danger of covert circularity, relying on a notion of the species in deciding the real limits of the reproductive community. In recent work, Eldredge has tended to shift from Mayr's original idea to the idea that species are defined by a shared "Species Mate Recognition System". But this seems to either recapitulate the same problems of bound, or, if these recognition systems are anatomically defined, we revert to the notion of specieshood defined by shared anatomical characters that Eldredge was in flight from. We revert to the notion of species as a class.

So species and avatars may have sufficient coherence to count as vehicles aiding the copying of high level replicators. I am much more doubtful about finding their phenotypes.

7. DO SPECIES HAVE PHENOTYPES?

If species are vehicles, they must have characteristics that aid replication. They must be adapted to aid replication. What might this involve?

Firstly, the phenotype must be built from irreducibly species properties, not properties of its members. But there is no clean criterion of a species-level property, though plausible examples include niche breadth and geographic range. These are not just species-level, or at least population level, properties, they are biologically important. Generalist species, and wide ranging species, are less apt to go extinct, and are less apt to speciate. But though range, for example, is not the property of any individual, it may be a simple and direct consequence of some individual morphological or behavioural trait. Consider the groups downstream from a splitting that vary in range. If that variance is the result of an individual difference (e.g., a nesting preference, or an anti-predator strategy), it seems reasonable to regard range as reducible in an interesting way to individual traits. The failure of one descendant population, and the success of the other, would not demonstrate that species are vehicles.

Second, the trait must be selectively relevant. There are irreducible properties, e.g. the total number of individuals in the history of the species. We would like to know if there are any that are causally salient to selection. But causal salience isn't enough. Individual properties can be relevant to the process of selection without being adaptations; for instance, the imperfection of genetic replication. If there were no mutation there would be less variation. But the propensity of genes to very occasionally mutate is not an adaptation; it's not something that is selected for. Similarly, population properties can be relevant to the process of selection without being adaptations. A lineage may be apt to split, for the individual organisms that compose it may be spread over, and isolated in, a large number of peripheral environments. The occupation of a large number of marginal environments may well yield a strong tendency for the gene pool to split. If so, the ranges a species occupies is a selectively relevant property of the species, but not an adaptation. It doesn't arise because that lineage outcompetes other lineages; indeed, the retreat to the margins is more likely to be a symptom of failure than success. Rather it's the source of variation. It, like the underlying rate of mutation, is certainly relevant to selection but isn't its creature. Species interaction requires species adaptations[16].

Moreover, added to the difficulties of reduction there is a problem about heritability. Let's consider, for instance, population density. Suppose an ancestral species splits into four daughters:

A x
B
C xxx
D xxx

Let's suppose C and D have higher population densities that A and B, and that they crash to extinction in a bad year, or that their denser population renders them vulnerable to disease. So they go extinct in virtue of a property of the population structure of the species. Suppose A is too thinly distributed over its range; it's just too hard to find a mate, and it expires too. Only the species of intermediate density B survives, and survives because of its population density. Prima facie, then, its population density is a species adaptation, and the species is genuinely an interactor. But reduction remains a problem. Once we have abandoned simple-minded pictures of reduction, population density is at least a candidate for reduction. Now, post splitting, the different population densities of A - D might well be due to a vast motley of interaction effects: there may be no salient individual properties of B that make the difference between B and the others. But in that case we have a problem about the mechanism of population regulation. I don't see how B could regulate its population (as it must if there is to be selection for a certain density) without there being a salient individual property which is the mechanism of that regulation. The demand that the property be irreducible cuts across the demand that the property be stable across generations and heritable by daughter species, characteristics a property must have to be an adaptation.

To see this, consider an example from a different context. In his [1982], Dawkins argues that we can profitably see natural selection as acting not on individual wasps, but on wasp strategies: digging a nesting tunnel versus invading an already built one. On plausible payoff assumptions, there is a mix of strategies at equilibrium: a digging frequency at which digging has the same average payoff as invading. At this point, it doesn't matter what the individual wasp does. In particular, there is no reason for it to dig at the equilibrium frequency, P*. It can always dig, or always invade, without average selective penalty. So we can see selection as operating in a frequency dependant way to stabilize the digging-invading proportion at around P*. The problem, analogous to population regulation, is one of mechanism. How is natural selection to stabilize that frequency except by building pure diggers/invaders at a certain ratio (analogous to a sex ratio) or by building wasps that dig at roughly P*, and invade at 1 - P*. So we can and should see selection operating on the wasps, not on strategies. I think this problem lies in wait for any putative species adaptation.

So its hard to see how a species could have a phenotype. First, as we have just seen, its hard for species to have traits that are both heritable and irreducible. Second, as Maynard Smith and others have argued, the subversion of high level processes by more powerful lower level ones makes the evolution of species adaptations for species replication most difficult. The generation time of individual organisms is swift compared to that of species or avatars. Sexually reproducing organisms exchange genetic material, whereas animal species rarely do. Moreover, there are many more organisms than species, so selection has more variation with which to work. So we can expect selection on organisms to be faster and more powerful than selection at higher levels, and to undercut high level adaptation building when the two forces are in conflict. Thirdly, as Dawkins as pointed out, the development of complex adaptation is tied to a developmental bottleneck, the growth of the new organism from a single cell. This developmental bottleneck ensures that organism level selection is not subverted at the selection at cellular level, for the replication fate of all the cell lineages funnels through the single germline. But still more importantly, Dawkins argues[17] that the initiation of a new developmental cycle is essential to the development of adaptation. A favourable change, by acting early in development, can change the whole organism, and can thus initiate the construction of something new. Reproduction "sets the switches back to zero" by beginning from a single cell. Literal beginning again from one cell is surely not necessary; there are intermediate cases like the caterpillar/butterfly transition. But species' adaptation building requires the new population to be especially plastic, if the species phenotype (not the phenotype of its individual members) is being rebuilt from near scratch. Gould and others have sometimes written that the forces maintaining stability break down at speciation events, but of course that plasticity is plasticity of organism phenotype, not species phenotype.

I do not think I have shown that we can rule out species adaptation, but it is certainly not easy to see how they could arise.

8. WHAT MIGHT SPECIES SELECTION EXPLAIN?

On the face of it, it is going to be hard to find examples of vehicle selection that resist redescription in individual terms. Certainly, directional changes in lineages do not look good candidates. Horses got bigger; hominids got bigger brained, relatively as well as absolutely. But overall size and brain size look to be paradigmatically properties of organisms, not species. So if there has been selection for these properties, the vehicles are organisms not species. That is true, surely, even if small horses found life tough not because of larger members of their own species, but because of larger horses from sibling species. If the directional pattern is the result of interspecific competition, that no more licenses us to talk of species selection than the human extermination of thylacines enables us to talk of species selection against marsupials[18]. If we are to find species selection, it will be on properties like population structure, niche breadth, or more recondite properties still.

It is hard to establish that an appeal to more recondite species level properties buys us explanatory power. In a recent paper, Schull[19] argues that species store information of such complexity, and use it so adaptively, that they are intelligent. Species are not just individuals, some are smart individuals. Now, if species are intelligent, different lineages will differ in intelligence. That's a difference on which species selection would act. We can expect intelligent lineages to be more diverse, or longer lived than their less bright rivals.

Schull argues that a species' intelligence consists in its capacity to respond to change, a capacity that he takes to be surprisingly well developed. He claims:

1. Species learn; we can see a species climbing a hill on an adaptive landscape as it acquiring and using information about its niche. Indeed we can see an organism's suite of adaptations to its niche as rich information about that niche. This information is a property of the species, not the individual organism. For the information held by any one organism is only a subset of the information in the whole gene pool. The species learns by modifying the information in the pool under the pressure of natural selection.

2. Hill climbing is a simple form of learning, but some species do better. Feral animals re-evolve wild traits quickly, for the wild traits are genetically masked, rather than eliminated, by domestic breeding. In these circumstances, a species (more exactly, an avatar) shows the benefit of previous experience. Just as sophisticated organisms will relearn a behaviour (or a slight modification of it) faster than they originally learned it, so the feral population learns much faster because it is unmasking memories, not creating new adaptations.

3. Species can be polymorphic, exhibiting distinct behaviours or morphologies in distinct environments. Aphids, for example, switch from asexual to sexual reproduction at the end of summer. Evolutionary biologists think of this as phenotypic plasticity. Schull urges us to see it as a species' capacity to follow conditional strategies: in circumstances A, do ___ ; in B, do ___ . This capacity in an organism, he reminds us, is a mark of some intelligence.

4. Schull buys into Wright's "shifting balance" theory of population structure to explain how species might escape local maxima; this too strikes him as a manifestation of a species' intelligence. For how do organisms gain insight; how do they come to rational expectations about what will happen in novel circumstances? They try trial and error in a simulated environment, thus avoiding the costs of error in their real environment by paying the costs of representing it.
Schull argues that the division of a species into many demes equally

allows a species to use trial and error without paying the full cost of error.

I think, perhaps, Schull is right in arguing that we can see species as individuals that process information. But I accept Lloyd Morgan's dictum that we should attribute intelligence only when no more parsimonious idea serves. I am unconvinced that we get explanatory mileage out of seeing species as intelligent. Consider just one of his examples: an information processing account of the anglerfish lineage. We can, sure enough, see the anglerfish lure as carrying information about the environment and prey of anglerfish. But what can we thus explain that is not equally well explained by an account of the natural selection of the individual fish or their genes? Nothing that I can see. Schull does not show that the attribution of information processing capacities to the species itself explains anything otherwise inexplicable. I suspect that it buys us nothing.

9. SPECIATION AS CAUSE AND EFFECT

In view of what has just been said, I am sceptical about the prospects of getting much mileage out of species selection. The fate of a species does seem to me to be an aggregation of the fate of its members; species selection buys us nothing. But I am inclined to think that speciation is a macroevolutionary phenomenon.

In two senses, speciation is irreducible to the change in the spread of individuals' traits in a population. First, and least dramatically, speciation is, I take it, multiply realized by changes at the individual level. By this I do not just mean that the details of the formation of a new species of horse differ from those of a new species of possum. Even when we abstract away from the details of horse and possum morphology variation in the speciating process remains. For example, the isolation of the speciating fragment can come about in many ways: geographic separation, a shift in mating season or timing of daily activity. Similarly, if, as Eldredge supposes, speciation is a change in the species mate recognition system, those too can change in a multitude of ways. Speciation, then, is multiply realized by individual level changes in body or behaviour. Those changes share nothing interesting at that level. But, more strongly, unless speciation just is the accumulation of a modest chunk of phenotypic change, it is invisible at the individual level. Inspection of the change in individual properties of organisms in a population over time won't tell you whether speciation has occurred. If you followed the lines of descent from a family of birds down the generations, you could tell a lot about the evolutionary change in that mini-lineage; about the changes in its struggles with predators and prey, about clutch size and plumage; about the size of range and territory. But you could not tell whether or where speciation had taken place; you could not tell where descendants had become a different species to their ancestors. That is particularly so if one buys into the hairy-chested version of cladism that regards all species downstream from a split as new species even if one daughter has identical traits to the ancestral species. In that case our family of birds can become members of a new species because, by great luck, a small flock blown out to sea three thousand miles away ended up on an island and diverged before going extinct[20]. It is in this sense that speciation is invisible at the level of individual history.

But is speciation no more than a macroevolutionary effect, an epiphenomenon of changes at the level of individuals in a population? Eldredge has argued that speciation has consequences for the course of evolution, an argument which in turn develops an idea of Mayr. He argues that speciation establishes barriers to the blending out of complex adaptation.

Eldredge has argued that adaptation is dependent on speciation. Individual selection does not proceed isolated from the processes through which species are born and die. Eldredge wants to rebut the idea that adaptive "economic" change[21] in the characteristics of individual organisms produces speciation. Some have taken such changes to produce speciation definitionally, supposing that the accrual of enough change is speciation. Others have supposed that economic change produces reproductive isolation, as changed-unchanged crosses cease to be viable, or as members of the diverging populations cease to regard one another as mates. Eldredge stands this view on its head: speciation consolidates "economic" change. The argument goes:

P1. Economic adaptation occurs initially in demes.

P2. Adaptations are not typically the result of a single genetic change, but of clusters of small genetic changes which must therefore be concentrated in particular individuals for the adaptation to be expressed in the phenotype.

but

P3. Demes are ephemeral

so

P4. Unless the change is somehow fixed, the death of the members of a defunct deme, or their integration into a new one, will lose the adaptation by backcrossing.

Hence

C. Speciation is essential to preserving economic change, by preventing migration/reintegration into the larger unadapted population.

In a slogan, No adaptation without speciation.

If this argument is right speciation is neither reducible nor epiphenomenal. It is a modest, but genuine, macroevolutionary cause. Not in any very exciting way; after all, every instance of speciation is identical to the change in the profile of individual organisms in particular populations. But a list of such changes would not suffice to explain what speciation was, or what its importance is. So there may be a modest place for macroevolution in evolutionary biology[22].

NOTES

[1]See for example Eldredge (1985a) *Time Frames,* Simon & Schuster, his (1985b) *The Unfinished Synthesis,* OUP and his (1989) *Macroevolutionary Dynamics,* McGraw Hill. See also Vrba & Eldredge (1984) 'Individuals,

Hierarchies and Processes: Towards a More Complete Evolutionary Theory' *Paleobiology* **10**, pp 146-171, Gould (1980d) 'Is a New and General Theory of Evolution Emerging?' *Paleobiology* , **6**, pp, 119-130, Gould (1983) 'The Meaning of Punctuated Equilibrium & Its Role in Validating a Hierarchical Approach to Macroevolution' *Scientia*, **1** pp135-157, Gould (1985) 'The Paradox of the First tier: An Agenda for Paleobiology' *Paleobiology* , **11**, pp 2-12, Salthe(1985) *Evolving Hierarchical Systems* ; Columbia UP, Stanley (1981) *The New Evolutionary Timetable* , Basic Books. For very sceptical views of the whole punctuational enterprise, see Dawkins (1986) *The Blind Watchmaker* ,W.W. Norton and Hoffman (1989) *Arguments on Evolution* , OUP.

[2] I discuss Gould on diversity in Sterelny 'Gould's 'Wonderful Life'', *Australasian Journal of Philosophy* , forthcoming.

[3] Gould especially; see for example his (1980b) 'The Episodic Nature of Evolutionary Change', and (1980c) 'The Return of the Hopeful Monster' both in Gould (1980a) *The Panda's Thumb* ; W.W. Norton.

[4] Sober (1984) *The Nature of Selection* , MIT Press (section 9.4) and Hoffman 1989 (chapter 7) have also made this crucial distinction between pattern and process in punctuated thought.

[5] Mayr seems to defend an intermediate position; the genetics of speciation is abnormal not because of macromutation but because founder populations are not just under greater selectional pressure, but also because they exhibit greater variability (cf Mayr (1988a) *Towards a New Philosophy of Biology* , Harvard UP, Cambridge pp 473-4).

[6] For a pellucid exposition of this, see Dawkins 1986 pp 66-74, but the general point is uncontroversial.

[7] See for example Mayr 1988a chapter 21.

[8] For a more detailed and careful reconciliation of adaptive hypotheses with due modesty about the optimizing power of natural selection see part I of Godfrey-Smith (forthcoming) *Teleometry and the Philosophy of*

Mind. Sober does not agree with this attempt to drive a wedge between optimality theory and adaptation; for his views on these issues see his (1987) 'What is Adaptationism?' in Dupre (1987) *The Latest on the Best*; MIT Press.

[9] This line of thought is particularly prominent in Eldredge; see his 1985 a&b, and his 1989.

[10] For the general distinction between the causally agnostic "selection of" some trait versus the causally committed "selection for" it, see Sober 1984a 3.2. For its application to species, see Vrba (1984) "What is Species Selection?" *Systematic Zoology* 33 pp 318-328.

[11] Hull (1988) 'Interactors versus Vehicles' in Plotkin (1988) *The Role of Behavior in Evolution* ; MIT Press and Dawkins (1982) *The Extended Phenotype* ; OUP, chapter 6.

[12] For ease of exposition I will take it that our candidate replicator is the species, not the species gene pool, for nothing I say depends on this distinction. For similar reasons, in this paper I will be neutral on the gene/organism debate.

[13] For example, in Part 4 of Maynard Smith (1989) *Did Darwin Get it Right?* , Chapman and Hall, and 9.3/9.4 of Sober 1984a.

[14] Some see these claims as tantamount to the idea that species are individuals; see Ghiselin (1974) 'A Radical Solution to the Species Problem' *Systematic Zoology* **23**, pp 536-544 , Hull (1978) 'A Matter of Individuality' *Philosophy of Science* **45** pp 335-360; reprinted in Sober (1984b) *Conceptual Issues In Evolutionary Biology* , MIT Press, and especially Hull (1987) 'Genealogical Actors in Ecological Roles' *Biology and Philosophy* , **2** pp 168-184; reprinted in Hull (1989) *The Metaphysics of Evolution* , State University of New York Press, New York. For arguments that we can pose the biological question independently of this issue see Kitcher (1989) 'Some Puzzles about Species' in Ruse *What the Philosophy of Biology Is* , Kluwer, and Hoffman 1989 chapter 8.

[15] See Kitcher 1989 on the idealizations implicit in the 'biological species concept'.

[16] See e. g. Williams (1966) *Adaptation and Natural Selection* , Princeton University Press, especially pp 96-98.

[17] In the final chapter of his 1982.

[18] See Sober 1984a pp 259-62 for a more detailed discussion of a parallel example drawing a similar moral.

[19] Schull (1990) 'Are Species Intelligent?' *Behavioral and Brain Sciences* **13** pp 63-108.

[20] See Ridley (1989) 'The Cladistic Solution to the Species Concept' *Biology & Philosophy* **4**, pp 1-16 for a defense of this picture of species and speciation.

[21] That is, changes in survivability and resource use, as contrasted with the sort of changes produced by "sexual selection".

[22] Thanks to John Maynard Smith for his comments on an very early version of this paper, and to Philip Kitcher, David Hull, and, especially, Elliot Sober for comments on the most recent version

Kim Sterelny
Department of Philosophy,
Victoria University of Wellington.

Robin Craw

MARGINS OF CLADISTICS: IDENTITY, DIFFERENCE AND PLACE IN THE EMERGENCE OF PHYLOGENETIC SYSTEMATICS, 1864 - 1975.

1. INTRODUCTION

"In conceptual evolution descent matters. If the history of taxonomy is to make any sense at all, who held a view and where he got this view is as important as what the view actually is."

David Hull, 1984

"It is doubtful that either Mitchell, Rosa or Hennig is the inventor of "cladistics" as a philosophy of classification for all, or almost all systematists seem to be, and seem ever to have been "cladists".

G. Nelson and N. Platnick, 1981

In 1929 the Russian emigre writer Vladimir Nabokov received the equivalent of US$8000 from contracts and newspaper articles. In his own words "he blew it all" on a butterfly collecting trip, his first in ten years. In February, 1930 Nabokov and his partner traveled south to the Pyrenees, collected Lepidoptera for 4 months, returning to Berlin in June. Nabokov took his specimens to be identified at the Deutsches Entomlogisches Institut at Dahlem on the outskirts of Berlin. In his paper describing this expedition he thanked Dr Walter Horn, the Director for his help.[1] But Nabokov was less than flattering about German entomological systematics. In his autobiography "Speak, Memory" he writes:

P. Griffiths (ed.), Trees of Life, 65–107.

"Great upheavals were taking place in the development of systematics. Since the middle of the [19th] century, Continental lepidopterology had been, on the whole, a simple and stable affair, smoothly run by the Germans. Its high priest, Dr Staudinger, was also the head of the largest firm of insect dealers. Even now, half a century after his death, German lepidopterists have not quite managed to shake off the hypnotic spell occasioned by his authority. He was still alive when his school began to lose ground as a scientific force in the world. While he and his followers stuck to specific and generic names sanctioned by long usage and were content to classify butterflies by characters visible to the naked eye, English-speaking authors were introducing nomenclatorial changes as a result of a strict application of the law of priority and taxonomic changes based on the microscopic study of organs. The Germans did their best to ignore the new trends and continued to cherish the philately-like side of entomology. Their solicitude for the 'average collector who should not be made to dissect' is comparable to the way nervous publishers pamper the 'average reader' who should not be made to think."[2]

Nabokov's poor opinion of German entomological systematics is idiosyncratic. Not only did German ornithological and entomological systematics of the period 1900-1940 contribute greatly to the formation of the evolutionary synthesis but major contributions to what has become known as cladistics/phylogenetic systematics were formuated within Central European systematics in the period immediately preceeding the Second World War.

The Deutsches Entomologches Institut was at that time one of the world's leading, if not the leading centre for entomological systematics. Important systematic works were published by this research institute in the 1920s-1930s, and an indication of the level of interest in evolutionary biology can be seen in the publication by Horn of Patrick Matthews' section on natural selection from his naval timbers work of 1831.[3] It was to this setting that Willi Hennig (1913-1976) came to work in 1937 as a fly specialist. Since his death in 1976 Hennig has been described as the "founding father of cladism", "the inventor of cladistics" and compared in stature with Darwin, Mendel, and Weismann.[4]

Cladistics/phylogenetic systematics is an approach to systematics where a natural classification equates to an estimation of evolutionary history, and is achieved by grouping taxa into single origin lineages (monophyletic groups) on the basis of their joint possession of shared derived characters (synapomorphies). Taxonomic groups based on shared primitive characters (symplesiomorphies) are termed paraphyletic and along with groups based on convergence (polyphyletic groups) have no place in a natural system.[5]

Cladistics and Willi Hennig, since his death in 1976, have come to prominence in the last decade for three reasons: (1) increasing numbers of systematists are accepting his method and theory of phylogenetic systematics in preference to the alternatives of numerical phenetics and evolutionary systematics; (2) testing hypotheses of evolutionary process in the context of phylogenetic hypotheses based on cladistic analyses has brought forth a reunification of systematics and evolutionary biology;[6] (3) David Hull has used the "cladistic revolution" as an empirical case study in his recent book "Science as a Process" in which he advanced a general evolutionary model for conceptual change.[7]

Hull has made a strong case for the use of actual examples in the philosophy of science. His basic argument concerning the acceptance of cladistic systematics is that due to language barriers Hennig's work remained obscure and little known outside Germay until through a fortuitous coincidence of circumstance it was widely disseminated and propagated in the systematics community by Dr Gareth Nelson of the American Museum of Natural History (New York) following the publication in 1966 of an English language text.[8] But inevitably inclusion of some examples leads to exclusion of others. Through the reconstruction of an alternative history of the development and dissemination of cladistics I will explore the critical silences and textual gaps in Hull's account. After establising the historical systematic context within which Hennig's phylogenetic systematics was developed and elaborated I will attempt to document the reception of his work inside the systematic community 1950-1975 through tabular summaries of the many detailed applications of his methodology by insect and other systematists, and the numerous citations of his works by systematists publishing on theoretical and methodological developments in the field. My focus is on mode of representation in diagrammatic form of the expression of the twin evolutionary processes of common ancestry and subsequent divergence in studies of the systematics of organisms.

2.THE TREE METAPHOR AND SYSTEMATICS

The discovery of the historicity of nature by Lamarck, Cuvier and Darwin constitutes a major rupture in the history of biology signified by the transformation of natural history into historical biology. Darwin, in 1837, was the first to introduce the tree metaphor into biology as a mode of re-presentation of the relationships of species in terms of common ancestry and subsequent divergence. The only figure in the "Origin" was a diagram of the genealogy of hypothetical taxa. Darwin never published an actual reconstruction of the phylogeny of a systematic group but he did privately work with such reconstructions as late as 1869.[9] The use of tree diagrams in systematics goes back long before Darwin but as Bowler has recently pointed out there are significant differences between pre- and post-Darwinian trees.[10] As well, the pre-Darwinian ideas as applied to phylogenetic trees have survived to this time, and are still being applied by some workers (Fig. 1).

3. THE SEARCH FOR A LOGICAL AND RATIONAL PHYLOGENETIC SYSTEMATICS IN CENTRAL EUROPE, 1864-1950.

Darwinism found more supporters amongst biologists and the general intellectual audience in Germany than in any other country. Invertebrate zoologists were particularly responsive to the explanatory power of the new evolutionary theory. Although earlier phylogenetic trees exist the earliest cladogram as such can be credited to the German carcinologist and entomologist Fritz Mueller, then living in Brazil. Mueller tested Darwin's ideas by applying them to the morphology, phylogeny and systematics of the Crustacea in his important book "Für Darwin" published in 1864.[11] Darwin described Mueller's book

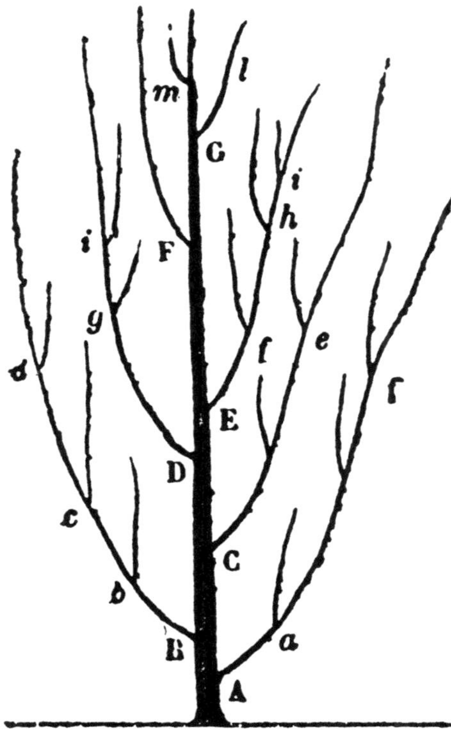

Figure 1A: Pre-Darwinian tree representing the devolopment of life from H.G. Bronn (1858; after Bowler, 1988).

Note the main stem, the sequencing of taxa and the pruned side brances.

"as remarkable" and it was considered so important by contemporaries that the Zoological Recorder for Crustacea, C. Spence-Bate devoted 9 pages to an extended abstract in English including a reproduction of the only tree figure in the book an analysis of relationships in the genus

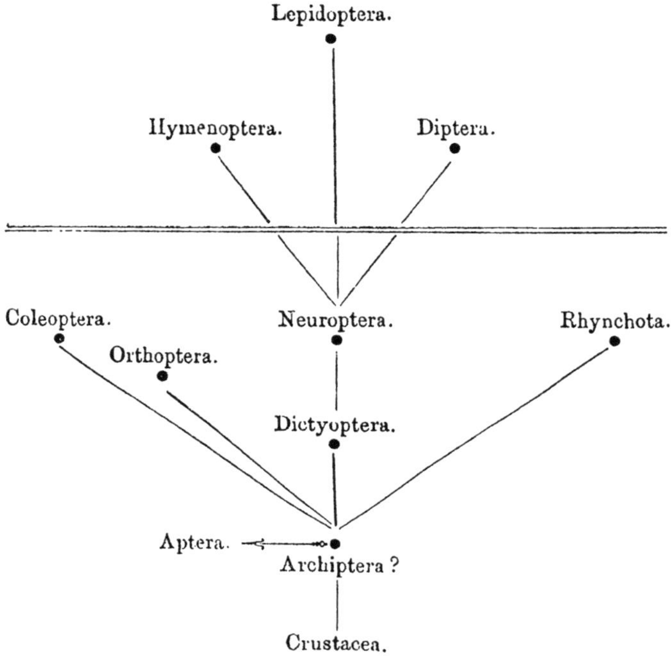

Figure 1B: Post-Darwinian phylogenetic tree which retains the main stem, the sequencing of taxa, and pruned side brances of pre-Darinian systematics.

The horizontal line divides the named taxa into two higher categories the Zygothoraca or Hohere Insekten [higher insects] (Hymenoptera, Lepidoptera and Diptera), and the Schizothoraca or Niedere Inskten [lower insects] (Coloptera, Orthoptera, Dictyoptera, Neuroptera and Rhynchota) neither of which is monophyletic. This tree also postulates that extant taxa (e.g. Dictyoptera, Neuroptera) are directly ancestral to other extant taxa (e.g. Lepidoptera, Diptera). (after G. Schoch, 1884, Uber die Gruppirung der Insekten-Ordnungen, *Mitt. Schweiz. Entomolog. Gesell.* 7: 34-36).

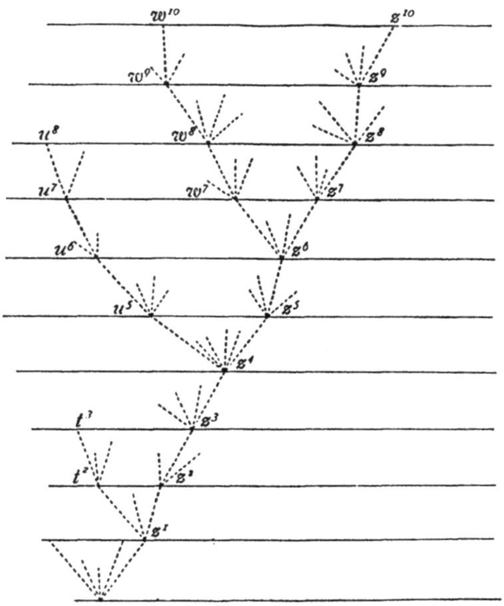

Figure 1C: Part of the orginal Darwinian phylogentic tree.

Note the lack of a main central stem and the divergently branched nature of the tree. Pruned side branches represent extinct taxa from earlier geological periods not extant taxa that are "lower" or more "primative" (after Darwin, 1859).

Melita.[12] This figure (Fig. 2) introduces the three taxon problem and its resolution in terms of conflicting characters which support alternative views of relationship. Mueller also divided characters into primitive and acquired (i.e. specialized, derived), and utilized ontogeny and outgroup comparsion as a means of polarising characters.

Numerous studies on the genealogical systematics of insects and other arthropods followed on from Mueller's work. By the 1870s-1880s major theoretical and practical studies on the phylogenetic

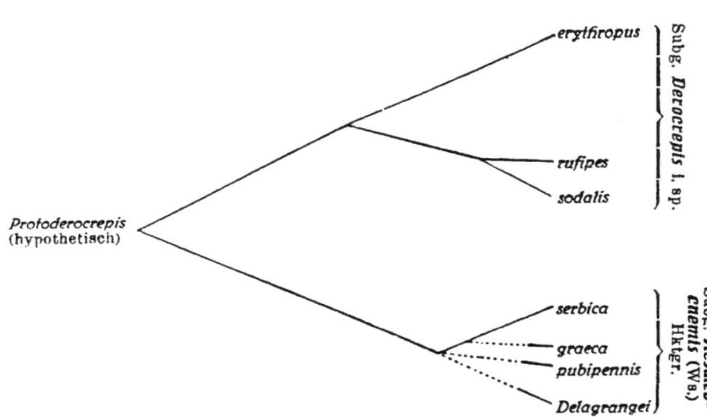

Figure 1D: Post-Darwinian phylogenetic tree after Heikertinger (1925).
Note the monophyletic descent from an hypothetical ancestral taxon (*Protoderocrepis*) and the terminus position accorded all extant taxa which are grouped into two monophyletic subgroups

systematics of insects had appeared. In a European context German and Austrian systematists had established themselves as leading theoreticians and practioners in the reconstruction of insect history by the turn of the century.[13] These fifty years of research into insect genealogy were summarized and extended by the Viennese entomologist Anton Handlirsch (1865-1935) who combined an extensive review and new descriptions of all fossil insects then known with work on the systematics of extant insects in his moumental monograph "Die Fossilen Insekten und die Phylogenie der Rezenten Formen" published in 1908. An important aspect of his study were phylogenetic diagrams illustrating the inter-relationships of all the orders and families of insects. Handlirsch's work is the perfect elaboration of Darwin's tree diagram as applied to insects; with his studies the two dimensional divergently branched tree with all extant taxa terminal becomes the paradigmatic mode of representation in insect phylogenetics.[14]

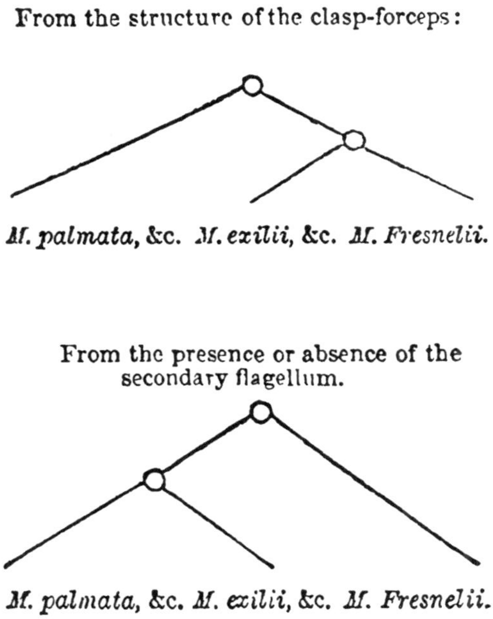

Figure 2A : Phylogenetic trees illustrating alternative views of relationship between species in the amphipod genus Melita after Mueller ([1863] 1869).

The tree on top based on the joint possession of the unique, specialized assymetric nature of the second pair of legs is preferred to the alternative tree based on the joint possession of a secondary flagellum, a character that is found in numerous other amphipod genera [i.e. outgroups comparison] and also appears early in ontogeny before disappearing in other taxa[i.e. ontogentic criterion].

By the 1920s-1930s the basic elements of what now constitutes cladistic/phylogenetic systematics were being applied in Central European botanical, molluscan, insect and avian systematics.[15] Hennig's earliest phylogenetic work on the systematics of the Tylidae, juxtaposing phylogenetic trees and geographical vicariance, resembles much

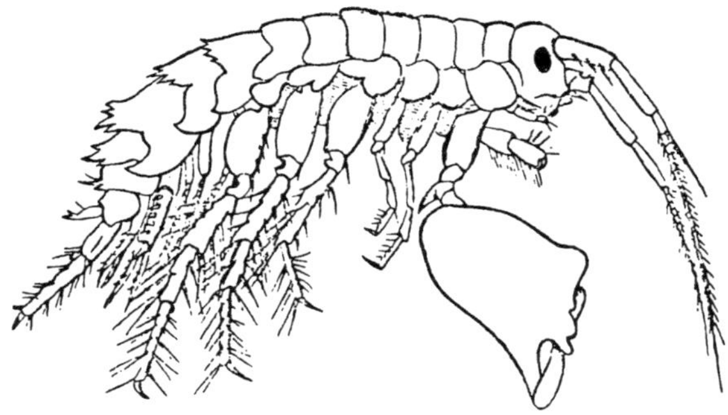

Figure 2B: Habitus side view of the amphipod *Melita exilii*.

Note the enormous second leg or clasp forcep, a unique derived character found in only five species of *Melita* which thus form a monophyletic group descended from a common ancestor (after Mueller [1863] 1869).

other work of this period.[16] In fact as Hennig later noted "In the years immediately preceding World War II the relationship between the so-called idealistic morphology and phylogenetics was debated particularly passionately in the German biological literature".[17] For example, at the 3rd German Entomological Conference held in 1929 a whole session was devoted to discussion of systematic philosophy.[18] General discussions of the relationship between natural classifications, phylogeny and systematics were published by several workers.[19] Innovations introduced include the mapping of geographic distributions and hostplants onto phylogenetic trees.[20]

Particularly important were extensive theoretical discussions and explicit methodological representations of the basic principles and applications of cladistic systematics by the German paleobotanist Walter Zimmerman (1892-1980)[21] and the Austrian ethologist Konrad Lorenz.[22] Works by both are cited in Hennig's later books and it is clear

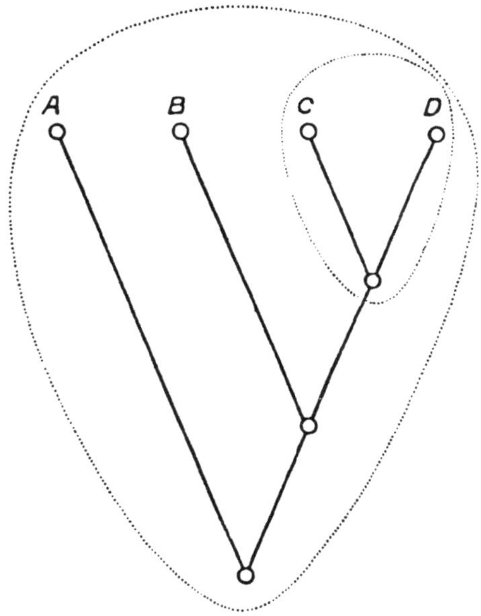

Figure 3A : Theoretical representation of nested monophyletic groups after Zimmermann (Zimmermann, 1931 : fig. 179, p.1004).

that much of what has been attributed to Hennig by later workers was actually derived from their discussions. Particularly important are Zimmermann's clear recognition of the principle of nested monophyletic groups and Lorenz's obvious anticipation of Hennig's argumentation scheme for phylogenetic analysis (Fig. 3). Zimmermann continued his explorations of the possibility of a phylogenetic systematics well into the 1960s. These explorations included an elegant discussion of the chain of being vs. the phylogenetic tree and an application of his methodology to a reconstruction of the phylogenetic relationships of genera in the buttercup family Ranunculaceae.[23]

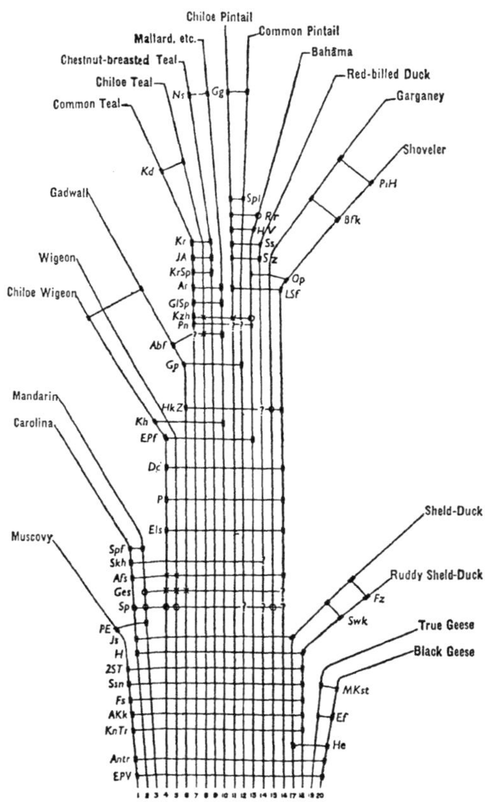

Figure 3B: Phylogenetic table Presenting the relationships of duck species of the subfamily Anatinae after Lorenz (1953: 90-91). (Reproduced by permission of the Avicultural Magazine.)

The vertical lines represent taxa (species) and the horizontal lines characters common to those species

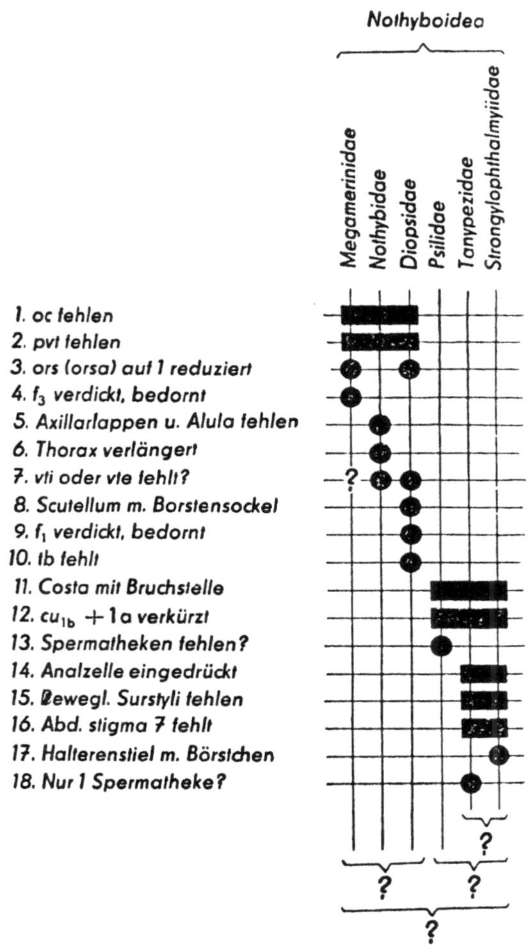

Figure 3C: Phylogenetic table or argumentation scheme representing the relationship between some fly families.

Vertical lines represent taxa (families) and horizontal lines with the heavy black bars characters common to those families (after Hennig, 1958 : Fig. 53, p563).

4. AN ITALIAN INTERLUDE: THE ANTICIPATION OF CLADISTIC SYSTEMATICS IN NORTHERN ITALY, 1890-1918

During the last decade of the 19th century an important group of theoretical biologists developed around the annelid specialist Daniele Rosa (1857-1944) at the Mueso di Zoologia dell'Universita di Torino in Turin, North-West Italy. This active group of zoological systematists, who made important contributions to the foundations of a distinctive Italian movement in evolutionary biology, included the orthopterist Giglio-Tos and the coleopterist Achille Griffini, both of whom published important books and papers on general biology. Yet neither Rosa nor his associates have been paid much, if any, attention by recent historians of biology despite the relevance of their work to such currently much debated topics as the status of species as historical entities, monophyly, the relationships between taxonomic grouping/ranking and phyletic branching, and the role of epigenetic mechanisms in evolution.

Indeed, Rosa appears to have anticipated many of the problems currently being discussed in evolutionary biology.[25] In particular Rosa can be credited with having made one of the first explicit and coherent formal statements of what has become the cladistic approach to systematics. Explicitly stated and discussed by Rosa in his 1918 book "Ologenesi"[26] are the following cladistic principles:

(1) natural groups/taxa should be strictly monophyletic;
(2) paraphyletic groups should not be admitted into a natural classification;
(3) the principle of dichotomous branching;
(4) the extinction of ancestral species after speciation;
(5) the rule of deviation;
(6) the principle of apomorphic and plesiomorphic branches.

Admitedly several of the above principles are no longer considered important components of cladistic systematists by some theorists but they were an important part of Hennig's original formulation of phylogenetic systematics although Hennig never cited Rosa's "Ologenesi". While some commentators have referred to these similarities between Hennig's and Rosa's work as "parallels" others have suggested that Hennig must have derived some of his ideas directly from Rosa's work.[27]

5. MEYRICK'S LAW, TUTT'S TREE AND MITCHELL'S THEOREM.

Open acceptance of Darwin's theory of the evolution of species did not lead automatically to acceptance of the divergently branched tree model or even to any visible change in systematic theory and practice amongst the new converts. Many traces of non-Darwinian developmental models of evolution can be found amongst the phylogenetic researches of such enthusiastic 'Darwinians' as E. Ray Lankester and T. H. Huxley.[28]

This tension between developmental and divergent models of evolution characterized the systematic work of Edward Meyrick who struggled throughout his scientific career to develop a natural classification of moths and butterflies that reflected "the natural genealogical order".[29] Meyrick's work also illustrates another recurrent theme in the history of systematics-the privileging and prioritising of "higher" vertebrate studies over those of "lower" invertebrates. Throughout the development of phylogenetic systematics several attempts have been made to formulate general principles or phylogenetic rules. The best known of these is the law of irreversibility or Dollo's law, named after its supposed discoverer the Belgian verterbrate palaeontologist Louis Dollo who first published it in 1893.[30] Almost ten years earlier Meyrick had already published this law:

"It seems to me useless to attempt to judge of the value of characters for classification, without strict reference to the principles of evolution. I think it might be laid down as an axiom, that when an organ has wholly disappeared in a genus other genera which originate as offshoots from this genus cannot regain the organ, although they might develop a substitute for it."[31]

The principle that phylogenetic reconstruction was the basis for a natural classification informed much late Victorian and Edwardian invertebrate systematics. Arthur Dendy, a sponge and flatworm systematist outlined the relationship between the Linnean hierarchy, the tree metaphor and phylogeny in his text "Outlines of Evolutionary Biology" which was widely used for teaching purposes in Britain and the British Empire during the early decades of this century.[32] Another lepidopterist J. W. Tutt attempted to formulate explicit principles for phylogenetic reconstruction clearly recognising the importance of tree thinking and common ancestry to systematic biology:

"The object of classification... is to place together those species which have most recently developed from the same stems; to work back, as far as may be, through the more recent stems to the less recent, and at last to that primeval form from which all have arisen. A system of classification, if it is to be a natural one, ought to be, when thoroughly worked out, a genealogical tree of the objects classified".[33]

Tutt applied these ideas in his studies on the relationships of lasiocampid moths and made theoretical contributions to genealogical systematics including an attempt at a phylogenetic definition of natural genera.[34]

The basic axiom of cladistics that monophyletic groups can only be recognized by the occurence of derived characters has been termed Mitchell's theorum in honour of P. Chalmers Mitchell, Secretary to the Zoological Society of London from 1903-1935, who published several expositions of this principle.[35] Mitchell called primitive character states archecentric and derived states apocentric, and argued that primitive character states were not evidence of close relationship or affinity:

"Likenesses which are due to the common possession of primitive features cannot be regarded as evidence of near relationship; that certain members of a group have retained what was once the property of all the members of that group can be no reason for placing such creatures close together in a system, if the system is to be based on blood-relationship. Resemblances that are due to the loss or reduction of parts that were once the property of the ancestral stock cannot afford a clear ground for inferring systematic affinity, since there is no difficulty in supposing the same organ to have been reduced or lost in independent branches of the same stock. Resemblances, on the otherhand, that are new acquisitions, that depend on definite anatomical peculiarities, must be the most likely field for the discovery of clues to affinity, simply because it appears to be less probable that the same anatomical device should have been produced independently than that it should have been acquired only once."[36]

While Mitchell's work had no immediate impact on systematic practice at least one ornithologist Percy Roycroft Lowe cited and discussed Mitchell's theorum in studies of the systematic relationships of the Madagascan rails and the ratite birds published in the 1920s.[37]

6. THE AUSTRAL IMPULSE: BIOGEOGRAPHY, CONTINENTAL DRIFT AND PHYLOGENETIC SYSTEMATICS

The history of Antipodean phylogenetic systematics begins in Dunedin, New Zealand on September 11th, 1883 with T. Jeffrey Parker's presentation of a phylogentic tree representing inter-relationships in the marine crayfish genus *Palinurus* to the Otago Institute. Although not the earliest tree representation to have labelled characters this is possibly the first tree to have characters inserted at the nodes (Fig. 4).[38]

Subsequent studies on the systematics of Southern Hemisphere organisms enabled Australasian and South American workers to establish unequivocable evidence of a close and direct phylogenetic relationship between now widely disjunct species and genera distributed upon the Southern continents and island groups by the early decades of this century.[39] Within this context Wegener's continental drift theory became accepted by many South African, Indian, South American and Australasian scientists as the explanation for the biological and geological similarities between Southern Hemisphere lands. While drift was not always advanced as an explanation of vicariant distributions by Southern Hemisphere biogeographers and systema-

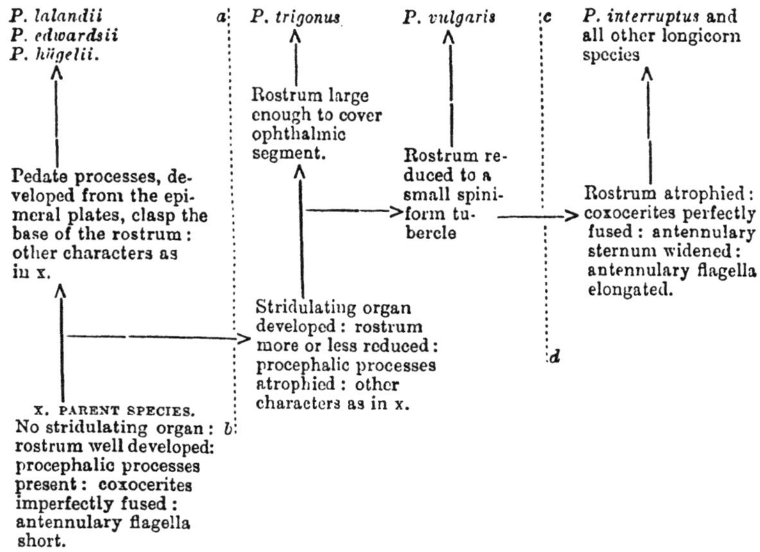

Figure 4: Phylogenetic tree/table of relationships amongst marine crayfish of the genus *Palinurus* after Parker, 1884 : 303.

Note the hypothetical common ancestor "X.Parent Species" and the labelling of ancestral nodes and linkages with characters. The vertical lines a-b and c-d represent suggested subdivisions of the genus. Note that the group to the right of a-b and the left of c-d (*P.trigonus* and *P.vulgaris*-subgenus *Palinurus*) is paraphyletic

tists some sort of Cretaceous - early Tertiary land connection between South America, Australia and New Zealand via the Scotia Arc and the Campbell Plateau around the South Pacific rim was generally accepted by the 1920s. By 1940 a comprehensive synthesis of the distribution of life under the process of continental rifting, drifting and collision had emerged for all the Southern Continents.[40]

The controversy generated by the implications for historical biogeography of Alfred Wegener's continental drift theory inspired a number of entomologists, especially Australasians, to examine closely the relationship between geographic distribution, phylogeny and systematics of the many insect groups disjunct between the Australasian region and South America. Two Australasian dipterists, G. H. Hardy and I. Mackerras published important studies in the early 1920s addressing such questions.[41] F. W. Edwards (1888-1940) of the British Museum, one of the leading fly systematists of the first half of the 20th century travelled to Southern Patagonia/Chile to collect Diptera in 1926 "inspired by a desire to test Wegener's hypothesis concerning continental drift, by means of a comparative study of the dipterous fauna of that region with those of Australia and New Zealand".[42] The results were published in a six volume set of 2,000 pages with contributions by all the leading fly specialists of the time.[43] Edwards also addressed the question of how phylogenetic relationships might be reconstructed:

"In estimating degrees of relationship therefore it is important, though not always easy, to distinguish between archaic or palingenetic and newer or coenogenetic characters; while among the latter we must further distinguish between those which are due to reduction and those which are accrescent or due to specialisation; and again among the accrescent features we must separate the adaptational, in the

	1953	1954	1960	1966	1969
Edwards	-	9	16	-	-
Hardy	-	7	2	-	-
Mackerras	-	1	7	-	-
Tillyard	12	6	1	1	47

Table 1: Number of citations of Australasian systematists and F. W. Edwards by W. Hennig, 1953 - 1969.

development of which environment as well as heredity has played a large part, from the non-adaptional, which are presumably due to the action of hereditary causes alone."[44]

Important as the contributions of Edwards, Hardy and Mackerras were to the development of Hennig's ideas as indicated by citations to their work by Hennig from 1953 onwards by far the major early inflence seems to have been the work of an Australian born entomologist Robin J. Tillyard (Table 1). Tillyard specialized in the phylogeny and classification of freshwater insects such as caddisflies, dragonflies, mayflies and stoneflies, and also made known the rich fossil insect fauna of the Queensland/New South Wales and Kansas Permian and Triassic. Tillyard anticipated many of the leading ideas of Hennig's phylogenetic systematics in his general principles of phylogenetic reconstruction. For instance in his 1917 study of dragonfly phylogeny Tillyard utilised a distinction between the primitive vs. the specialized state of characters, and the outgroup and ontogenetic criteria for polarising characters in order to construct a dichotomous and divergently branched phylogenetic diagram. That Tillyard utilised what is basically a cladistic method is quite apparent from his 1921 study of stonefly classification in which taxa are grouped on the basis of shared derived caracters and the data arranged in a character by taxa matrix (Fig. 5).[45]

Tillyard directly inspired a number of systematists to undertake phylogenetic annalyses of the groups they were revising. Amongst those influenced were the New Zealand Hemipterist J. G. Myers who mapped ecological habitats against phylogeny in his 1929 revision of New Zealand cicadas[46] and the young English entomologist Cyril L. Withycombe who corresponded with Tillyard about the systematics of the alderflies (Neuroptera).[47] Withycombe's major innovation was to displace the privileged position fossils had hitherto occupied in phylogenetic reconstruction:

"With reference to palaeontological findings, it is well known that such data, as all other, may at times be misleading. If fossils of a particular form have not been found it merely shows that, if existent, the habits, environment, structure, etc., of this form were not such as to favour

TABLE SHOWING PRINCIPAL CHARACTERS FOR THE FAMILIES OF THE ORDER PERLARIA.

Character	Family						
	Eustheniidæ	*Austroperlidæ*	*Pteronarcidæ*	*Perlidæ*	*Leptoperlidæ*	*Capniidæ*	*Nemouridæ*
(1) Mandibles:—A normal; B, reduced to lamina	A	A	B	B	A	A	A
(2) Clypeus and Labrum:—A, normal; B, hidden	A	A	A	BU	A	A	A
(3) Palpi:—A, with short joints; B, one or more joints elongated	A	A	A	B*	A	B*	B*
(4) Anterior coxæ:—A, wide apart; B, approximated	A	A	BU	A	A	A	A
(5) Tarsal joints: — A, 2 least, 3 longer than 1; B, otherwise	A	A	A	B†	A	B†	B†
(6) Cerci:—A, with 5 or more joints; B, reduced to a single joint	A	A	A	A	A	A	BU
(7) Outer margin of hind-wing:—A, complete convex whole; B, with re-entrant angle at distal end of Cu_2	AU	B	B	B	B	B	B
(8) Anal fan:—A, with cross-veins; B without	AU	B	B	B	B	B	B
(9) Cross-veins in distal half of forewing: — A, present; B, absent	A	A	A	A(B)	A	B	B.
(10) Cubito-anal cross-veins in forewing: — A, present; B, absent	AU	B	B	B	B	B	B
(11) Branches of Rs in fore-wing:—A, 3 or more; B, 2 or 1	A	A	A	A(B)	B	B	B
(12) Branches of Cu_1 in fore-wing:—A, 3 or more; B, 2 or 1	A(B)	A(B)	A	A(B)	B	B	B
(13) Anastomosis or transverse cord:—A, absent; B, present	A	A	A	A(B)	A	B	B
(14) 1A in forewing: — A, present; B, absent	A	A	A(B)	B	B	B	B
(15) 3A in forewing: — A, forked; B, simple	A(B)	A	A	A(B)	A(B)	B	A
(16) Primitive paired lateral gills on abdomen: — A, present on segs. 1–5 or 1–6; B, absent	AU	B	B‡	B‡	B‡	B	B‡
Percentage of archaic characters§ for the most archaic members of each family	100	75	63	44	56	25	25

Figure 5A: Table illustrating the distribution of archaic (primitive) (A) and specialised (derived) (B) characters amongst families of stoneflies (Plecoptera) (after Tillyard, 1921: p.7)

fossilisation. On the other hand, when insects of a particular family are found in older rocks than those of any other family in the same order it does not mean to say that the insects found in the oldest strata are

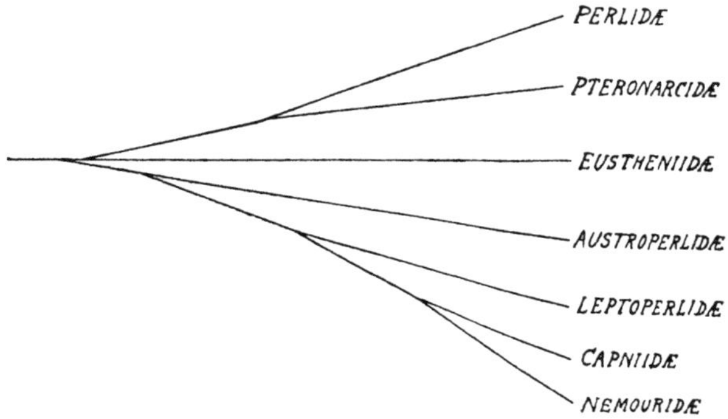

Figure 5B: Phylogenetic diagram showing the relationships of the families of stoneflies (Plecoptera) (after Tillyard, 1921: p.7).

necessarily more ancient types than many fossilised at a later date, and certainly one is not justified in assuming them as ancestral to present-day forms."[48]

7. AMERICAN PHYLOGENETIC SYSTEMATICS: AN INDEPENDENT PARALLEL DEVELOPMENT?

Like their German counterparts, several American entomologists become enthusiastic supporters of evolutionary ideas from the 1860s onwards. By 1900 a number of important papers on the genealogical reationships of insect orders and families had been published although they all retain elements of pre-Darwinian evolutionary theory and the chain of being due to their authors adherence to orthogenetic theories of evolution.[49]

From 1900 onwards a steady stream of papers on insect phylogenetics were published by American workers.[50] Often these papers contain general theoretical discussions of phylogenetic system-

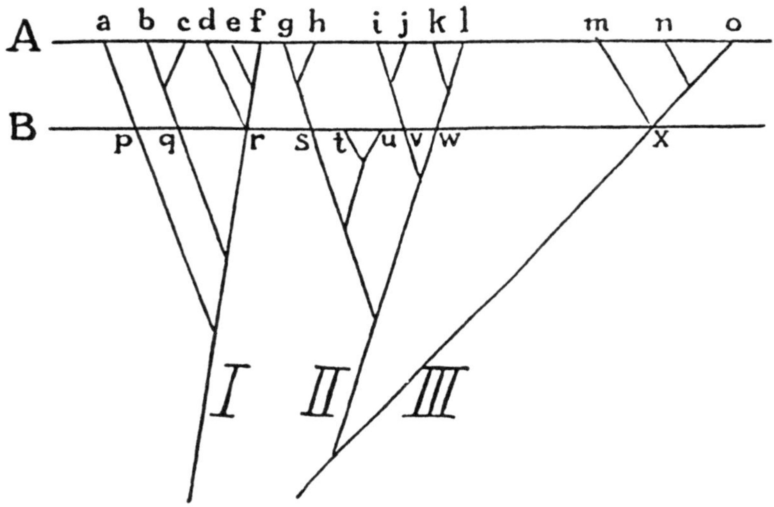

Figure 6: Diagramatic phylogenetic trees after Busck (1909 : Fig. 5, p.101).

atics of which the work of August Busck ranks as the most sophisticated in its recognition that phenetic similarity and level of organisation (grade) do not necessarily equate to close phylogenetic relationship:

"We have been doing our classification too much horizontally, so to say- any twig on the entire phylogenetic tree which has reached a certain type of imaginal structure has been placed in such or such a genus or family without sufficient regard for its origin. This does not produce a natural system.

In the above diagramatic phylogenic tree it is of course the easiest to say that everything which has reached a certain level A is to be considered a systematic entity and that which reaches to the level B is another family of subfamily, as the characters may warrant, irrespective of whether it originated from main branch I, II, or III.

> But such a system would be purely mechanical and not according to nature's own divisions. And while we with our limited powers of observation are forced to adopt to some extent some mechanical arrangement in order to get any classification at all, it is manifestly obligatory, when we can trace the phylogeny and realize the true natural lines, that we utilize such knowledge and not continue our temporary mechanical system, merely because it is easier.
>
> Thus in the above illustration it is clearly correct to place the species (or genera) a, b, c, d, e, and f in the same family or genus A as their common characters may warrant, but in this family must not be placed g, h, i, ... to o." [see Fig. 6][51]

Other important developments and innovations in American phylogenetic systematics during the early decades of this century include quantitative tabular comparisons of primitive versus derived character states amongst taxa, the recognition of congruence between host-parasite phylogenies, and the use of behavioural characters in phylogenetic reconstruction.[52] The microlepidopterist Annette Braun clearly distinguished between primitive, derived and unique adaptive (divergent) characters in her recognition that the moth family Micropterygidae was more closely related to the rest of the Lepidoptera than to the caddisflies (Trichoptera) where a number of studies had placed them.[53]

In the early 1920s the herpetologist Charles L. Camp introduced a number of criteria for determining phylogeny that anticipate the principes of cladistics.[54] Entomologist and arachnologist Joseph Conrad Chamberlain attempted to correlate categorical rank, geological horizon and the branching process in his studies of arthropod phylogeny.[55] In the early 1930s W. T. M. Forbes, a lepidopterist began to develop a quantitative methodology for deriving ancestor-descendent and divergent phylogenies.[56]

Against this background of extensive methodological progress Herbert Ross (1908-1978) independently anticipated Hennig's methodology in his 1937 study of sawfly classification and phylogeny. In later studies Ross continued to refine his approach in important studies on general insect phylogenetics, and caddisfly systematics that are extensively cited by Hennig. Ross also anticipated contemporary historical ecology suggesting a synthesis of ecology, evolutionary theory and phylogenetic systematics as early as 1956, and in 1974 he published the first American textbook containing an explicit methodological approach to phylogenetic reconstruction.[57] As well Ross influenced a number of students to undertake studies on the phylogenetic systematics of many insect groups.[58]

8. THE RECEPTION OF HENNIG'S WORK IN THE SYSTEMATIC COMMUNITY

The story of the acceptance and application of Hennig's methodology in the Americas begins not in New York at the American Museum of Natural History but at the Fundacion "Miguel Lillo" in Tucuman, Argentina where two European War refugees Aczel and Kusenov came to work as insect systematists in the late 1940s. Aczel was working on a revision of Latin American flies of the family Tylidae upon which group Hennig had published the major systematic and biogeographic monograph to date, including the first explicit application of his phylogenetic systematics. Aczel's study published in 1951 is the first known application of Hennig's methodology by another author.[59] Petr Wygodzinsky, a colleague of Aczel's in the 1950s, introduced Hennig's phylogenetic systematics to scientists at the American Mseum of Natural History when he moved to New York in 1962.[60] Wygodzinsky had also given a seminar on phylogenetic systematics at the Berkley campus of the University of California in April 1961 while there on a fellowship.[61]

By the mid-1950s American, Australian and South African fly systematists had begun discussing Hennig's phylogenetic work in considerable depth (Table 2) so that by 1963 one of them could write:

"Hennig . . .has adopted a different approach. He believes that phylogeny, in the strict sense of lineage, should be given absolute priority in making a classification. . .This system has a firm foundation, in that it accepts speciation. . . as the point of seperation of lineages, and the hierarchy of classification that is constructed is completely consistent with the hierachy of lineage. It . . .seems to be finding favour, because one finds workers on various groups beginning to use 'apomorphic' and 'plesiomorphic' almost as everyday terms."[62]

Hennig's 1960 paper on the dipterous fauna of New Zealand demonstrated that the existence of a close relationship between taxa on the southern landmasses was not sufficient evidence for a former direct land connection between the areas. What was needed as unequivocable evidence were detailed patterns of repeated phylogenetic relationships.[63]

This paper was particuarly influential in spreading Hennig's ideas amongst Australasian workers. L. J. (Jack) Dumbleton, a New Zealand systematist exchaged reprints with Hennig in the early 1960s. In 1963 he arranged for a translation of Hennig's 1960 paper and distributed copies to 14 universities, museums and government scientific institutions in New Zealand, and to five individuals including the two Australians E. Marks and I. Mackerras.[64]

In Europe Hennig's ideas had been widely disseminated, accepted and applied by a number of invertebrate systematists by 1960, and by even more workers after that date (Tables 2, 3 & 4). That Hennig's ideas had been widely discussed, especially in entomological circles prior to 1960 is evident by various presentations at the Tenth International Congress of Entomology held in Montreal in August, 1956;[65] frequent citation in major review papers in the late 1950s[66] and a major presentation to the Zoological Institute of the Soviet Academy of Sciences by Hennig's director in 1958.[67] By the mid-1960s extensive discussions and applications of Hennig's work took place in spaces

Table 2: (opposite) : Examples of General Citation and Discussion of Hennig's Phylogenetic Systematics Work, 1950 - 1972.

A plus sign after a date in this table, and tables 3 and 4 indicates that the named worker published further papers in later years citing and discussing Hennig's phylogentic systematics. Full Reference details for the works listed in tables 2,3 and 4 can be found in *Biological Abstracts* and the *Zoological Record*. Two references which may prove elusive are T. Borgmeier, 1950, [review of] Hennig: Die Larvenformen der Dipteren (2.Teil), *Rev. de Entomogia* **21**:690-691, and G. Kuschel, 1952, [review of] Willi Hennig, Die Larvenformen der Dipteren, *Rev.Chil. Ent.* **2**:319.

Worker	Date	Country
(1) Europe		
Kiriakoff	1952+	Belgium
Gunther	1953+	East Germany
Szelenyi	1955	Hungary
Colosi	1956+	Italy
Tortonese	1957	Italy
van Emden	1957	England
Theowald	1958	Netherlands
Boettger	1958	East Germany
Hinton	1958	England
Knulle	1959	East Germany
Tollet	1959	Belgium
Prinicis	1960	Sweden
Bahrman	1960	East Gemany
Illies	1960+	West Germany
Rohdendorf	1961	U.S.S.R.
Crowson	1961+	Scotland
Stammer	1961	Austria
Lindeburg	1962+	Finland
Giles	1963	England
Abdullah	1964+	England
Gisin	1964	Switzerland
Sharov	1965+	U.S.S.R.
Tuomikoski	1966+	Finland
Lubischchev	1966	U.S.S.R
Mickoleit	1966+	West Germany
Tobias	1967	U.S.S.R.
de Lattin	1967	East Germany
Kaufman	1967	U.S.S.R
Povolny	1967	Czechoslovalia
Kuhn-Schnyder	1967	Switzerland
Wickler	1967	West Germany
Vane-Wright	1967+	England
Janetscek	1967	Austria
Wagner	1967	West Germany
Grasshoff	1968	West Germany
Reiss	1968	West Germany
Kuhne	1968	West Germany
Speight	1969	England
Hackman	1969	West Germany
Besch	1969	West Germany
Banarescu	1970+	Roumania
Puthz	1971	West Germany
de Viedma	1972	Spain
(2) North and Central America		
Bequaert	1953+	U.S.A.
Sabrosky	1961+	U.S.A.
Simpson	1961+	U.S.A
Sokal	1962+	U.S.A
Lopez-Ochoterena	1964	Mexico
Johnston	1964	Canada
Boyes	1964	Canada
Mayr	1965+	U.S.A.
Mettrick	1965	Jamaica
Douwnes	1968	Canada
Throckmorton	1968	U.S.A.
Corliss	1968	U.S.A.
Spenser	1969	Canada
Vockeroth	1969	Canada
Gosline	1969	U.S.A.
Marx/Rabb	1970	U.S.A.
(3) Southern Hemisphere		
Borgmeir	1950	Brazil
Aczel	1951+	Argentina
Kusenov	1951	Argentina
Kuschel	1952	Chile
Mackerras	1954+	Australia
Hardy	1955+	Australia
Zumpt	1957+	South Africa
Wygodzinsky/Coscaron	1962+	Argentina
Dumbleton	1963	New Zealand
McAlpine	1966	Australia
Watt	1968	New Zealand
Rapoport	1968	Argentina
Craig	1969	New Zealand
Dugdale	1969	New Zealand
Guimaraes	1972	Brazil

Worker	Date	Country
Brundin	1956+	Sweden
Stuckenberg	1958	South Africa
Karl	1959	East Germany
Strenzke	1960	Finland
Steyskal	1961	U.S.A.
McAlpine	1963+	Canada
Griffiths	1964+	England
Panelius	1965	Finland
Marks	1968	Australia
Serra-Tosio	1968	France
Thompson	1969	U.S.A.
Lehrer	1970	Roumania

Table 3: Examples of Explicit Detailed Applications of Hennig's Methodology in Dipteran Systematics 1956 - 1970.

and places as diverse and different as Kingston, Jamaica and Helsinki, Finland giving rise to a number of distinct cladistic research schools (Tables 2, 3 & 4). Particularly prominent were Diptera systematic research groups at the Biosystematics Research Institute in Ottawa (Canada) where Hennig spent three months in 1968,[68] and the USDA in Washington D.C. where Carl Sabrosky could date personal contact with Hennig back to their joint attendance at the 7th International Congress of Entomology held in Berlin in 1938.[69] The earliest distinct group of personally inter-linked cladistic researchers were Hennig's colleagues at the D.E.I. (e.g. Machatschke) and friends in East German museums (e.g. Guenther) that extended to workers in groups other than insects by the early 1960s. Also by this time a distinctive West German group of researchers in phylogenetic systematics had emerged centred on the Max Planck Institut fur Limonologie.[70]

A loosely based group emerged in Scandinavia originating from the early application of Hennig's method to the systematics of midge flies by Lars Brundin, of the Rijksmuseum in Stockholm, in 1956.[71] Hennig's ideas spread rapidly to other Scandinavian entomologists particularly Finnish systematists at the Zoological Museum in Helsinki who not only applied the methodology in their own studies of organ-

Worker	Date	Country	Taxonomic Group
Machatschke	1954+	East Germany	Coleoptera
Gunther	1955+	East Germany	Orthoptera
Paclt	1956	Czechoslovakia	Protura
Koningsman	1960	East Germany	Pthiraptera
Friese	1960	East Germany	Lepidoptera
Oeser	1961	East Germany	Plecoptera
Illies	1961+	West Germany	Plecoptera
Nowakowski	1962	Poland	Lepidoptera
Besch	1963	West Germany	Acari
Karg	1965	East Germany	Acari
Wygodzinsky	1966	U.S.A.	Hemiptera
Kristensen	1967+	Denmark	Lepidoptera
Evers	1968	West Germany	Coleoptera
Zwick	1968+	West Germany	Coleoptera
Gaskin	1968	New Zealand	Lepidoptera
Koponen	1968+	Finland	Musci
Schlee	1969+	West Germany	Hemiptera
Ball/Erwin	1969+	Canada	Coleoptera
Whitehead	1969+	Canada	Coleoptera
G. Nelson	1969+	U.S.A.	Vertebrates
Van der Hammen	1970	Netherlands	Acari
Herman	1970+	U.S.A.	Plecoptera
Szelegiewicz	1971	Poland	Hemiptera
Nelson/Hanson	1971	U.S.A.	Plecoptera
Yasuda	1972	Japan	Lepidoptera
Smithers	1972	Australia	Pscoptera
Meinander	1972	Finland	Neuroptera

Table 4: Examples of Explicit Application of Hennig's Methodology in Non-Dipteran Systematics, 1954 -1972.

isms but also made significantly different theoretical contributions that continue to this day, including attempts at synthesizing Hennig's phylogenetics with Mayr/Simpson evolutionary systematics.[72]

Central to this wide diffusion of Hennig's ideas was the fact that there was not prior to 1965 only one discussion of his approach -the formidable 1950 book of 370 pages-as is commonly believed but a number of clearer, conciser and simpler expositions of cladistics.

These include methodological introductions to detailed accounts of insect and fly phylogenetic systematics, an influential general review paper and the study of the New Zealand Diptera.[73] Hull and other writers on the acceptance of cladistics caim that the 1966 book of Hennig's in English had a major impact. Study of the 1965-1975 Science Citation Index entries for Hennig does not indicate this at all. Between 1965-1969 there were 146 citations to publications by Hennig of which 10 are to the English language review article of 1965 and 14 to the 1966 book while there are 51 citations to the earlier German theoretical and methodological literature of which only 8 are to the 1950 book. The English language publications of Hennig began to become more cited in the period 1970-1975 (Table 5) slowly overtaking citation of German language items but there remained a significant number of the latter still cited.

9. THE DIFFERENCE THAT SPACE MAKES

In contrast to Hull's overdeveloped historico-biographical contextualization of scientific social life and productivity centred on

Date	**Citations**	
	Earlier German Work (1948 - 1960)	**English Language** (1965 - 1966)
1970	15	23
1971	16	19
1972	7	13
1973	11	29
1974	13	29
1975	12	21
Totals	**74**	**134**

Table 5: Number of Citations of Hennig's Theoretical and Methodological Work, 1970 - 1975. [Source: Science Citation Index]

the notion of an innovative ideological elite in New York I have proposed an explicitly spatialized narrative concerning the multiplicity of places and spaces wherein phylogenetic systematics began and developed, and finally emerged as cladistics. In contrast Hull's historicist mode of narration imposed a straight story of phylogenetics sequentially unfolding in time from Darwin through Matthew to Simpson, Mayr, Brundin and Nelson. Traversing that storyline laterally produces a shift from sequence to simultaneity, historicity to spatiality, biography to geography.

(Bio) geographical analysis of scientific change reamains underdeveloped. But the social relations of scientific production and reproduction are both space-forming and space-contigent. The difference that place makes to the constitution of scientific social relations remains problematic. Phylogenetic systematics offers a site, a place, a space that many contenders attempt to appropriate. The conceptual lineages of phylogenetic systematics leading to cladistics were not divergent branches cleaved off a single trunk or centre but rather braided rivers cast in a diversity of geo-cultural spaces. Hull's social evolutionary theory of science rationalizes existing conditions within one scientific marketplace and may merely serve to promote repetitive behaviour.

ACKNOWLEDGEMENTS

I am grateful to J. S. Dugdale, Gareth Nelson and Robert O'Hara for useful comments on an earlier version of this paper, and to Paul Griffiths for editorial assistance and patience.

NOTES

[1] V. Nabokov, 1931, Notes on the Lepidoptera of the Pyrenees orientales and the Ariege. *Entomologist* **64**: 255-257, 268-271. Nabokov's trip and his relationship with staff at the Deutches Entomologisches Institut are described in A. Field, 1986, *VN The Life and Art of Vladimir Nabokov*, Crown Publishers Inc., New York and B. Boyd, 1990, *Vladimir Nabokov:*

the Russian Years, Chatto and Windus, London.

[2] V. Nabokov, 1957, Butterflies, pp. 18-19 in P. Matthews (ed.), *The Pursuit of Moths and Butterflies,* Chatto and Windus, London.

[3] W. Horn, 1936, Introduction to P. Matthew "On naval timber and Arboriculture", *Arb. Morph. taxon. Ent.* **3**: 233-237. Details of Walter Horn's life and work can be found in R. Korschefsky, 1939, Dr. Walter Harn, *Ent. Blatter* **35**: 177-184, and A. Sachtleben, 1939, Dr. Walter Horn zum Gedachtnis, *Arb. Morph. tax. Ent.* **6**: 201-222. The early history of the Deutsches Etomologisches Institut is summarized by Anon., 1930, 25 Jahre Deutsches Entomologists Institut, *Wien. Ent. Zeit.* **47**: 60-61.

[4] See e.g., C. Dupuis, 1984, Willi Hennig's impact on taxonomic thought, *Ann. Rev. Ecol. Syst.* **15**: 1-24 and E. Sober, 1988, *Reconstructing the Past: Parsimony, Evolution annd Inference.* MIT Press, Cambridge. Details of Willi Hennig's life, career and bibliography can be found in G. W. Byers, 1977, In memoriam Willi Hennig (1913-1976), *Jl. Kansas Ent. Soc.* **50**: 272-274; D. Schlee, 1978, In memoriam Willi Hennig 1913-1976. Einne biographische Skizze, *Entomological Germ.* **4(3-4)**: 377-391, and Anon., 1978, In memoriam Willi Hennig [bibliography only], *Beitr. Ent.* **28**: 169-177. An important and insightful study of the recent history of cladistics in Europe and North America has been provided by C. Dupuis, 1979, Permanence et actualité de la Systématique: La «Systématique phyogénétique» de W. Hennig (Historique, discussion, choix de références), *Cahiers des Naturalistes, N.S.* **34**: 1-69.

[5] Standard accounts of phylogenetic systematics/cladistics are W. Hennig, 1966, *Phylogenetic Systematics,* University of Illinois Press, Urbana; G. Nelson and N. Platnick, 1981, *Systematics and Biogeography: cladistics and vicariance,* Columbia Univeristy Press, New York; E. O. Wiley, 1981, *Phylogenetics: The Theory and Practice of Phylogenetic Systematics,* John Wiley and Sons, New York; and N. Eldredge and J. Cracraft, 1980, *Phylogenetic Patterns and the Evolutionary Process,* Columbia University Press, New York.

[6] See e.g., V.A. Funk and D. R. Brooks, 1990, Phylogenetic Systematics

as the basis of Comparative Biology, *Smithsonian Contributions to Botany* **73.**

[7] D. Hull, 1988, *Science as a Process,* The Universty of Chicago Press, Chicago and London.

[8] See e.g., D. Hull, ibid, pp. 130-131: "Science is supposed to be international, and to some extent it is, but language differences can form very real barriers. In 1950, an East German entomologist Willi Hennig published a formidable treatise entitled "Grundzuge einer Theorie der Phylogenetischen Systematik". In this work Hennig took seriously the claim that phylogenetic classifications are to represent phylogeny.

"Initially Hennig had little impact on taxonomic disputes among English speaking systematists", and p. 363: "Nelson initiated the Hennigian revolution among English speaking systematists" and p. xi: "While I was studying the pheneticists, I realized that Nelson, without intending to, had given rise to a new "school" -the cladists or phylogeneticists".

So Hull claims a prominent role for Nelson in the late 1960s to early 1970s as the major disseminator of cladistics. These events are fortuitously connected to publication of the English language book by Hennig in 1966.

[9] See, e.g. G. De Beer, (ed.) 1960, Darwin's notebooks on the transmutation of Species Part 1. First Notebook (July 1837 - February 1838). *Bull. Brit. Mus. (Nat. Hist.) Historical series* **2 (2)**; C. Darwin, 1859, *On the origin of species*. John Murray, London and "Taxonomical tree showing hypothetical relation of man to other primates "drawn by Darwin on April 21, 1868 figured in S. S. Schweber, 1989, John Herschel and Charles Darwin: A study in parallel lives, *Jl. Hist. Biol.* **22 (1)**: 1-71.

[10] Pre- and post-Darwinian tree diagrams have been illustrated and discussed by H. J. Lam, 1936, Phylogenetic symbols, past and present (Being an apology for Genealogical trees), *Acta Biotheoretica* **2**: 153-194; G. Nelson and N. Platniick, 1981, op. cit.; R. J. O'Hara, 1988, Diagramatic classifications of birds, 1819-1901: views of the natural system in 19th-

century British ornithology. pp. 2746-2759 in *Acta XIX Congress Internationalis Ornithologici* (H. Ouellet, ed.) and 1990 Representations of the natural system in the nineteenth century, *Biology and Philosophy* **6**: 255-274; and P. F. Stevens, 1983, Augustun Augier's "Arbre botanique" (1801), a remarkable early botanical representation of the natural system, *Taxon* **32**: 203-211. The theoretical differences signified by pre- and post-Darwinian trees are discussed in P. J. Bowler, 1989, *The non-Darwinian revolution: reinterpreting a historical myth,* The John Hopkins University Press, Baltimore and London.

[11] F. Mueller, 1864, *Für Darwin,* Leipzig [English translation 1869, *Facts and Arguments for Darwin,* John Murray, London]. Mueller is best known for his discovery of the phenomenon known as Muellerian mimicry.

[12] C. Spence-Bate 1865, Crustacea, pp. 257-311 in *The record of Zoological Literature, Volume the First.* John van Voorst, London.

[13] See e.g., A. Dohrn 1867, Eugeron Boeckingi und die Genealogie der Arthropoden, *Stett. Ent. Zeit., 1867:* 145-153; A. Speyer 1870, Zur genealogie der Schmetterlinge, *Stett. Ent. Zeit., 1870:* 202-223; P. Mayer, 1876, Ueber Ontogenie und Phylogenie der insekten, *Jenaische Zeit. fur Naturwissenschraft* **10**: 125-221; F. Brauer, 1885, Systematisch-zoologische Studien, *Sitzb. der. kais. Akad. der Wissensch.* **XCI**: 237-384 and 1887, *Beziehungen der Descendenzlehre zur Systematik,* Adolf Holzhausen, Wein; C. Boerner, 1904, Zur systematik der Hexapoden, *Zool. Anz.* **27**: 511-533; and P. Speiser, 1908, Die geographische Verbreitung der Diptera pupiparia und ihre Phylogenie, *Zeit. fur. wiss. Instekten.* **13**: 241-246, 301-305, 420-427, 437-447.

[14] A. Handlirsch, 1908, *Die Fossilen Insekten und die Phylogenie der rezenten formen,* Wilhelm Englemann, Leipzig. The standard account of Handlirsch's life including a bibliography is by M. Beier, 1935, Anton Handlirsch, *Konowia* **14**: 340-347.

[15] See e.g.; *F. Heikertinger, 1925, Monographie der Halticinengattung Derocrepis Weise* (Coleopt,. Chrysomelidae), *Wein. Ent. Zeit.* **42**: 95-178;

H. Hoffman, 1929, Zur Kenntnis der Oncidiien (Gastrop. pulmon.). Ein Beitrag zur geographischen Vebreitung, Phylogenie und Systematik dieser Familie. II Tiel. Phylogenie und Verbreitung. *Zool. Jahrb. Jena Abt. Syst.* **57**: 253-302; K. Eller, 1939, Fragen und Probleme zur Zoogeographie und zur Rassen- und Artbildung in der *Papilio machaon*-Gruppe, *Verh. VII Int. Kog. für Ent.* **1**: 74-101; E. Fischer, 1937, Der Basaldorn der Schmetterlinge und seine phylogenetische Bedeutung, *Ent. Zeit.* **50**: 290-295; G. Save-Soderbergh, 1934, Some points of view concerning the evolution of the vertebrates and the classification of the group, *Arkiv fur Zoologi* **26 (A)**: 1-20; and H. V. Boetticher, 1943, Die phylogenetisch-systematische Stellung von Anseranas, *Zool. Anz.* **142**: 55-58

[16] W. Hennig, 1934, Revision der Tyliden (Dipt., Acalypt.). *Stett. Ent. Zeit.* **95**: 65-108, 294-330; 1935, Revision der Tyliden (Dipt., Acalypt.). II. Teit. *Konowia* **14**: 68-93, 192-216, 289-310; 1936, Beziehungen zwischen geographischer Verbreitug und systematischer Gliederung bei einigen Dipterenfamilien:ein Beitrag zum Problem der Gliederung systematischer Kategorien hoherer Ordnung. *Zool. Anz. 116:* 161-175.

[17] W. Hennig, 1966, *Phylogenetic systematics*, University of Illinois Press, Urbana, p. 9.

[18] F. van Emden and W. Horn (eds.) 1929 **3.** *Wanderversammlung Deutscher Entomologen in Giessen.* Berlin-Dahlem.

[19] See e.g., F. Ruschkamp, 1927, Systematik und Stammesgeschichte, *Ent. Mitt.* **16**: 420-422.

[20] Geographic distributions of taxa are mapped onto a phylogenetic tree by G. Grimpe and H. Hoffman, 1925, Die Nacktschnecken von Neu-Caledonien, den Loyalty-Inseln und den Neuen-hebriden, *Nova Caledonica A. Zoologie,* **3**: 339-476, and gymnosperm hosts are mapped onto a phylogenetic tree of their associated rust fungi by P. Dietel, 1938, Betrachtungen zur Entwicklung des Stammbaums der Pucciniastreen, *Annales Mycologici* **36**: *1-8.*

[21] W. Zimmermann 1931, Arbeitsweise der botanischen Phylogenetik, pp. 941-1053 in E. Abderhalden (ed), *Handbuch der biologischen Arbeitsmethoden. Abt. IX, Teil 3.*

[22] K. Lorenz, 1941, Vergleichende Bewegungsstudien an Anatinen, *Jl. fur Ornith., Erganzungsband* III: 194-293; and 1953, Comparative studies on the behaviour of Anatinae, *Avicultural Magazine* **59**: 80-91.

[23] W. Zimmermann, 1962, Kritische Beitrage zu Einigen Biologischen Problemen. IV. Die Ursachen der Evolution, *Acta Biotheoretica* **14**: 121-206, and 1965, Familie Ranunculaceae, In Hegi, *Illustrierte Flora von Mitteleuropa.* Bd. III/3,2 Aufl. -Munchen.

[24] The most comprehensive account of Rosa's work is G. Colosi, 1961, L'Opera di Daniele Rosa e la dottrina dell'evoluzione (con cenni biografici e bibliografici). *Mem. Acad. Sci. Torino Serie 3, Tomo* **4**: 329-368.

[25] The relevance of Rosa's work has been discussed by G. Nelson, 1973, Coments on Leon Croizat's biogeography, *Syst. Zol.* **22**: 312-319 and 1974, Historical biogeography: an alternative formalization, *Ibid* **23**: 555-558; C. Baroni Urbani, 1977, Hologenesis, phylogenetic systematics and evolution, *Ibid* **26**: 343-346; 1979, The causes of evolution: converging orthodoxy and heresay, *Ibid* **28**: 622-624; 1990, Searching for the evolutionary roots of cladistics (A simplified conspectus of Hologenetic theory), *Newsletter of the OSAKA GROUP for the study of Dynamic Structures, Nov. 1990:* 1-8, and R. Craw and M. Heads 1988, Reading Croizat: on the edge of biology, *Riv. Biol./Biology Forum* **81 (4)**: 499-532.

[26] D. Rosa 1918 *Ologenesi: Nuova teoria dell'evoluzione e della distrituzione geografica dei viventi.* Bemporad and Figilio, Firenze.

[27] The similiarity between Hennig's and Rosa's views has been discussed by G. Colosi, 1956, Filogensi e Sistematica, *Boll. Zool.* **23**: 787-824; G. Nelson and N. Platnick, 1981, *op. cit.*, pp. 325-236; and L. Croizat, 1975, *Biogeografia analitica y sintetica ("Panbiogeografia") de las Americas,*

Tomo II, Biblioteca de la Academia de Ciencias Fisicas, Mathematicas y Naturales, pp. 609-613, 824-828.

[28] Peter J. Bowler, 1989, Development and Adaptation: Evolutionary Concepts in British Morphology, 1870-1914. *Brit. Jl. Hist. Sci.* **22**: 283-297.

[29] E. Meyrick, 1985, *A Handbook of the British Lepidoptera*, MacMillan and Co., London, p. 13. Meyrick's scientific work is discussed in Gaden S. Robinson, 1986, Edward Meyrick: an unpublished essay on phylogeny. *Jl. Nat. Hist.* **20**: 359-367.

[30] Dollo's law is discussed in S. J. Gould. 1970, Dollo on Dollo's law: Irreversibility and the Status of Evolutionary Law's, *Jl. Hist. Bio.* **3**: 189-212.

[31] E. Meyrick, 1884, On the classification of the Australian Pyralidina. *Trans. Ent. Soc. Lond., 1884*, p. 277. A more extended treatment by the same worker is 1889, On the interpretation of neural structure, *Entomol. Mon. Mag.* **25**: 175-178. Meyrick's priority over Dollo was noted by H. Sachtleben, 1951, Zur Prioritat des Satzes von der Irreversibilitat der Entwicklung, *Bet. Ent.* **1**: 93.

[32] A. Dendy, 1912, *Outlines of Evolutionary Biology*, Constable and Co. Ltd., London.

[33] J. W. Tutt, 1895, An attempt to correlate the results arrived at in recent papers on the classification of Lepidoptera, *Trans. Ent. Soc. Lond., 1895*, p. 343.

[34] J. W. Tutt, 1898, Some considerations of natural genera, and incidental references to the nature of species, *Proc. South Lond. Ent. Nat. Hist. Soc., 1898:* 20-30, and 1899, The Lasiocampids, *Proc. South Lond. Ent. Nat. Hist. Soc., 1899:* 1-11.

[35] See e.g., P. Chalmers Mitchell, 1901, On the intestinal tract of birds, with remarks on the valuation and nomenclature of zoological char-

acters, *Trans. Linn. Soc. Lond., Zool., ser. 2*, **8**: 173-275. The term Mitchell's Theorum was coined by R. Carolin, 1984, Mitchell's Theorum and its impact on biology, *Cladistics, Systematics and Phylogeny Symposium, Canberra, Abstracts*, p. 1.

[36] P. Chalmers Mitchell, 1905, On the intestinal tract of mammals, *Trans. Zool. Soc. Lond.*, **17**, pp. 528-529.

[37] P. R. Lowe, 1924, On the anatomy and systematic position of the Madagascan bird *Mesites (Mesoaenas)*, *Proc. Zool. Soc. Lond., 1924:* 1131-1152 and 1928, Studies and observations bearing on the phylogeny of the Ostrich and its allies, *Proc. Zool. Soc. Lond., 1928:* 185-247.

[38] T. J. Parker, 1883, On the structure of the head in *Palinurus*, with special reference to the classification of the genus, *N. Z. Jl. Sci.* **1**: 584-585.

[39] See e.g.; F. W. Hutton, 1884, On the origin of the fauna and flora in New Zealand, *Ann. Mag. Nat. Hist.* **(5) 13**: 425-448.

[40] See e.g. L. Harrison, 1928, The composition and origins of the Australian fauna, with special reference to the Wegener hypothesis, *Report 18t meeting ANZAAS*, pp. 332-396, and L. King, 1944, On palaeogeography, *S. Afr. Geogr. Jl.* **26**: 1-13.

[41] See e.g.; G. H. Hardy, 1922, The geographical distribution of genera belonging to the Diptera Brachyera of Australia, *Aust. Zool.* **2**: 143-147 and I. M. Mackerras, 1925, The Nemestrinidae (Diptera) of the Australasian region, *Proc. Linn. Soc. N.S.W.* **50**: 489-561, and 1927, Notes on Australian mosquitoes (Diptera, Culiculidae). Part ii. the zoogeography of the subgenus *Ochlerotatus*, with notes on the species, *Proc. Linn. Soc. N.S.W.* **52**: 284-298. This work is reviewed in G. H. Hardy, 1951, Theories of the World distribution of Diptera, *Ent. Mon. Mag.* **87**: 99-102.

[42] N. D. Riley, 1940, Dr. F. W. Edwards, F.R.S., *Nature 146:* p. 740. See also F. W. Edwards, 1928, An account of a collecting trip to Patagonia and Southern Chile, *Proc. 4th Int. Congr. Ent.*: 416-417.

[43] See e.g.; C. P. Alexander, 1929, *Diptera of Patagonia and South Chile. Part 1-Crane-flies.* British Museum (Natural History), London.

[44] F. W. Edwards, 1926, The phylogeny of Nematocerous Diptera: a critical review of some recent suggestions. *Proc. 3rd Int. Congr. Ent.* **2**: 114-115.

[45] Details of Tillyard's life and work can be found in A. D. Imms, 1938, Robin John Tillyard, 1881-1937, *Obituary Notices of the Royal Society of London* **2(6)**: 339-345; J. W. Evans, 1946, Robin John Tillyard, 1881-1937. *Proc. Linn.. Soc. N.S.W.* **71**: 252-256; and T. K. Crosby, 1977, Robin John Tillyard-the man behind the book, *N. Z. Ent.* **6(3)**: 305-308. Tillyard's approach to phylogenetic reconstruction and polarization of character states is most comprehensively covered in R. J. Tillyard, 1917, A study of the rectal breathing-apparatus in the larvae of Anisopterid dragonflies. *Jl. Linn. Soc. (Zoo.)* **33**: 127-196; 1918, The Panorpid Complex: A study of the phylogeny of the Holometabolous Insects, with special reference to the subclasses Panorpoidea and Neuropteroidea [Introduction], *Proc. Linn. Soc. N.S.W.* **43**: 265-284; and 1921, A new classification of the order Perlaria, *Canad. Ent.* **53**: 35-43.

[46] J. G. Myers, 1929, The taxaonomy, phylogeny and distribution of New Zealand cicadas (Homoptera), *Trans. Ent. Soc. Lond., 1929:* 29-60.

[47] Withycombe's correspondence with Tillyard is noted in Anon., 1927, Cyril Luckes Withycombe, M.Sc., Phd., *Ent. Mon. Mag.* **63**: 16-17.

[48] C. L. Withycombe, 1924, Some aspects of the biology and morphology of the Neuroptera. With special reference to the immature stages and their possible phylogenetic significance, *Trans. Ent. Soc. Lond.*, 1924: 403.

[49] See e.g.; A. S. Packard, 1883, The genealogy of Insects, *Amer. Nat.*, 1883: 932; H. Osborn, 1895, The phylogeny of Hemiptera, *Proc. Ent. Soc. Wash.* **3**: 185-190; W. H. Ashmead, 1896, The phylogeny of the Hymneoptera, *Ibid.* **3**: 323-336; and A. Radcliffe Grote, 1897, An attempt to classify the

Holarctic Lepidoptera by means of specialization of the wings, *Jl. N.Y. Ent. Soc.* **5**: 151-160.

[50] See e.g.; H. Dyar, 1901, Life history of *Callidapteryx dryopterata* Grt., *Proc. Ent. Soc. Wash.* **6**: 414-418; H. Osborn, 1908, The habits of insects as a factor in classification, *Ann. Ent. Soc. Amer. 1:* 70-84; C. L. Turner, 1916, Breeding habits of the Orthoptera, *Ann. Ent. Soc. Amer.* **9**: 117-135; J. A. Hyslop, 1917, The phylogeny of the Elateridae, *Ann. Ent. Soc. Amer.* **10**: 241-263; E. M. Walker, 1922, The terminal structures of orthopteroid insects: A phylogenetic study, *Ann. Ent. Soc. Amer.* **15**: 1-89; H. E. Ewing, 1922, The phylogeny of the gall mites and a new classification of the suborder Prostigmata of the order Acarina, *Ann. Ent. Soc. Amer.* **15**: 213-222, and H. Good, 1925, Wing venation of the Buprestidae, *Ann. Ent. Soc. Amer.* **28**: 251-276.

[51] A. Busck, 1909, Notes on Microlepidoptera, with descriptions of new North American species, *Proc. Ent. Soc. Wash.* **11**: 87-103.

[52] See. e.g., H. Osborn, 1908, *op. cit.;* C. L. Turner, 1916, *op. cit.;* C. W. Leng, 1920, *Catalogue of the Coleoptera of American, North of Mexico,* J. D. Shermman, New York, [see pp. 7-25]; and M. H. Hatch, 1925, The phylogeny and phylogenetic tendencies of Gyrinidae, *Pap. Mich. Acad. Sci. Arts. Letters* **5**: 429-467.

[53] A. Braun, 1919, Wing structure of Lepidoptera and the phylogenetic and taxonomic value of certain persistant trichopterous characters, *Ann. Ent. Soc. Amer.* **12**: 349-366. Annette Braun's studies on phylogenetic systematics and her anticipation of contemporary cladistics have been reviewed by M. Alma Solis, Annette Frances Braun: Early concepts in Lepidoptera phylogenetics, *Am. Entomol.* **36**: 122-126.

[54] S. M Moody, 1985, Charles L. Camp and his 1923 Classification of Lizards: an early cladist?, *Syst. Zool.* **34**: 216-222.

[55] J. C. Chamberlain, 1923, A systematic monograph of the Tachardiinae or lac insects, *Bull. Ent. Res.* **14**: 147-212.

[56] W. T. M. Forbes, 1933, A grouping of the argotine genera, *Ent. Amer.* **14**: 1-30; 1936, The classification of the Tyatiridae,*Ann. Ent. Soc. Amer.* **29**: 779-803; and 1939, Revisional notes on the Danainae(Lepidoptera), *Ent. Amer.* **19**: 101-139. The similarities between Forbes' method and contemporary cladistics has been discussed by P. R. Ackery and R. I. Vane-Wright, 1984, *Milkweed butterflies: their cladistics and biology*, Dept. of Entomology, British Museum (Natural History), London.

[57] Short accounts of Ross' life are given by G. Byers, 1978, In memory of Herbert H. Ross, 1908-1978, *Jl. Kansas Ent. Soc.* **52 (1)**: 92, 108, and J.C. Morse and R.T. Allen, 1979, Herbert Holdsworth Ross, *Syst. Zool.* **28**: 413-414. His more important works are H. H. Ross, 1937, A generic classification of the Nearctic sawflies (Hymenoptera, Symphyta). *Illinois Biological Mongraphs* **15(2)**; 1956, *Evolution and Classification of the Mountain Caddisflies*, University of llinois Press, Urbana; 1958, The relationships of systematics and the principles of organic evolution, *Proc. 10th Int. Congr. Ent.* **1**: 423-429; and 1974, *Biological Systematics*, Addison Wesley Pub. Co., Inc. Mass.

[58] See e.g.; S. Kramer, 1950, The morphology and phylogeny of Auchenorhynchous Homoptera (Insecta), *Ill. Bio. Mon.* **20(4)**; S. S. Roback, 1951, A classfication of the muscoid calyptrate Diptera, *Ann. Ent. Soc. Amer.* **44**: 327-361; L. J. Stanard, 1957, The phylogeny and classification of the North American genera of the suborder Tubulifera (Thysanoptera), *Ill. Bio. Mon.* **25**; E. L. Mockford, 1965, The genus *Caecilus* (Psocoptera : Caeciliidae). Part 1. Species groups and the North American species of the falvidus group. *Trans. Amer. Ent. Soc.* **91**: 121-166; and R. T. Allen, 1972, A revision of the genus *Loxandrus* LeConte (Coleoptera : Carabidae) in North America, *Ent. Amer.* **46**: 1-184.

[59] M. L. Aczel, 1951, Morfologia externa y Division Sistematica de las "Tanypezidiformes", *Acta Zoo. Lilloana* **11**: 483-589. Hennig is mentioned by N. Kusnezov, 1951, El genero *"Pogonomyrmex"* Mayr (Hym., Formicidae), *Acta Zoo Lilloana* **11**: 227-333.

[60] R. T. Schuh and . H. Herman, 1988, Petr Wolfgang Wygodzinsky (1916-1987), *J. New York Ent. Soc.* **96**: 227-232.

[61] This seminar is noted in J. A. Slater and J. T. Polhemus, 1990, Peter D. Ashlock 1929-1989, *J. New York Ent. Soc.* **98**: 113-118.

[62] I. M. Mackerras, 1964, The classification of animals, *Proc. Linn. Soc. N.S.W.* **88**: 324-335.

[63] W. Hennig, 1960, Die Dippteren-Fauna von Neuseeland als systematisches und tiergeographisches Problem, *Beit. Ent.* **10**: 221-329. [Translated by P. Wygodzinsky as 1966, The Diptera fauna of New Zealand as a problem in systematics and zoogeography, by Willi Hennig, *Pacific Insects Monogr.* **9**: 1-81.

[64] Papers and letters inserted inside translation of Hennig, 1960, DSIR Plant Protection reprint collection no. 35046.

[65] See e.g. E. Handschin, 1958, Die systematische Stellung der Collembolen, *Proc. 10th Int. Congr. Ent.* **1**: 499-516.

[66] See e.g.; F. I. van Emden, 1957, The taxonomic significance of the characters of immature insects, *Ann. Rev. Ent.* **2**: 91-106; H. E. Hinton, 1958, The phylogeny of the panorpoid orders *Ibid. 3:* 181-206 and J. L. Gressitt, 1958, Zoogeography of insects, *Ibid.* **3**: 182-230.

[67] H. Sachtleben, 1958, The activity of the German Entomological Institute of the German Academy of Agricultural Sciences in Berlin, *Ent. Rev.* **37**: 665-670.

[68] W. Hennig, 1969, Neue Gattungen Arten der Acalyptratae (Diptera:Cyclorrhapha), *Canad. Ent.* **101**: 589-633.

[69] C. W. Sabrosky, 1978, The family position of the peculiar genus *Horaismoptera* (Diptera:Tethinidae), *Ent. Germ.* **4**: 327-336. Sabrosky's and Hennig's attendance at the 7th International Congress of Entomology, Berlin, 21-26 August, 1938 is noted in *Verh. Vll Int. Kong. für Ent.* **5** : pp XLIII, LXIV.

[70] This group of workers included W. Besch, J. Illies and P. Zwick. This group appears to have influenced at least one English speaking systematist W.D. Williams to adopt Hennig's phylogenetic systematics, see the acknowledgements and the citation of Hennig's "Grundzüge" in W.D. Williams, 1970, A revision of North American Epigean Species of *Asellus* (Custacea : Isopoda), *Smithsonian Contributions to Zoology*, **49** : 74-77.

[71] L. Brundin, 1956, Zur systematik der Orthocladiinae (Dipt., Chironomidae), *Rep. Inst. Freshwat. Res. Drottningholm, t.* **37**: 1-185.

[72] See e.g.; Lindeberg, B. 1962, The abdominal spiracles in Chironomidae (Diptera) with some notes on the phylogeny of the family, *Ann. Ent. Soc. Fenn.* **28**: 1-10, and R. Tuomikoski, 1966, Generic taxonomy of the Exechiini (Dipt., Mycetophildae), *Ann. Ent. Soc. Fenn.* **32**: 159-194. A synthesis of Hennig's phylogenetic systematics and Mayr/Simpson evolutionary systematics is suggested by S. Panelius, 1965, A revision of the European gall midges of the Porricondylinae (Diptera, Insecta), *Acta Zollogica Fennica* **113**.

[73] see e.g., W. Hennig, 1953, Kritische Bermerkungen zum phylogenetischen System der Insekten, *Beit. Ent.* **3** : 1-85, and 1957, Systematik und Phylogenese, *Bericht ub. Hundertjahrfeier Dtsch. ent. Gesellsch.*, Berlin, pp 50-71.

Robin Craw,
DSIR Plant Protection Unit,
Auckland.

SECTION TWO

CENTRAL CONCEPTS OF EVOLUTIONARY THEORY

Paul Griffiths

ADAPTIVE EXPLANATION AND THE CONCEPT OF A VESTIGE

1. INTRODUCTION

The debate over 'adaptationism' concerns the *extent* to which the morphology and behaviour of animals can be understood as adaptations to the environment. Some traits are pretty clearly adaptations, and others are pretty clearly not. The question is one of proportion. This paper makes a number of related points about adaptive or functional explanation in evolutionary biology. Overall, I try to show that adaptive and functional explanations have been unfairly caricatured in recent attacks on 'adaptationism'. I hope that by clarifying the nature of adaptive explanation I can make clear some of the empirical issues that need to be settled before 'adaptationism' can be sensibly evaluated.

In section one of the paper, I describe the connection between adaptive explanation and the ascription of biological function. In section two I consider Gould and Vrba's[1] adaptation/exaptation distinction and their related function/effect distinction. Using the account of biological function outlined in section one, I criticise both distinctions, and substitute new ones. The new distinctions allow a better understanding of adaptive explanation. Given certain empirical presuppositions, the prospects for adaptive explanation are not as bad as Gould and Vrba make them seem.

In section three I discuss the way in which explanations of the acquisition of biological function have been confused with explanations of the acquisition of the traits which bear these functions. Once again, given this clarification, and certain empirical presuppositions, things look better for adaptive explanation. In the final section, I

P. Griffiths (ed.), Trees of Life, 111–131.

discuss the loss of biological function, and concept of a vestige. The distinctions and presuppositions discussed in previous sections allow us to give a clear characterisation of a vestigial trait. This is important, because such a characterisation is an essential requirement of any theory of biological teleology.

2. FUNCTIONS

Biological functions are teleological. The functions of a thing is what that thing is *for*. Some biologists find this embarassing, as evinced by the euphemism 'teleonomy'. They should not be embarassed. Biological teleology can be made perfectly scientifically respectable.

The teleological nature of biological functions is revealed by the connection between function ascription and explanation. To ascribe a biological function is to commit oneself to providing an explanation in which the function figures. Thus, for example, someone who claims that a behaviour is a mechanism for controlling the population is immediately challenged to provide a account of the origin of the behaviours which avoids implausible group selective mechanisms[2]. The sense of 'function' in which functions support functional explanations has come to be referred to in the philosophical literature as 'proper function'.

There are many other senses in which an object may be said to have a function. Many of these have legitimate uses in science[3]. These senses of function are distinguished from proper functions because it is not legitimate to demand a functional explanation to accompany them. The function of the liver in certain disease processes is as a harbour for parasites, but the form and prevalence of the liver cannot be explained by reference to its function in this sense. Such senses of function are not teleological.

Correct ascriptions of proper, teleological function are correct, must be backed by valid functional explanations . So if biological functions are proper functions there must be valid biological explanations in which these functions figure. Suitable explanations were readily available when it was believed that biological systems were divinely created. Functional explanations worked by pointing out the creators intentions. In post-Darwinian biology, however, the ascription of proper functions is made possible by the idea that some traits of biological objects are *adaptations*. The prevalence and form of these traits can be explained by the fact that they enhanced the fitness of ancestors who bore them. If a trait is not an adaptation it may be said to perform certain functions, in a sense that has no teleological implications, but it has no proper function. Thus, to take G.C Williams' classic example, the fact that flying fish are heavier than air may function as a device to return them to the water, but this is not one of its biological functions, because its physical inevitability makes an adaptive explanation unnecessary. To use the standard terminology, this is an effect of weight, not a function.

The adaptive justification of biological teleology has been expanded by philosophers into the 'etiological theory' of proper functions[4]. According to the etiological theory, the proper functions of a trait are those effects of earlier examples of the trait which caused the trait to be reproduced. If a trait has such effects then the current prevalence of the trait can be partly explained by the capacity of the trait to perform this function. The teleological properties of proper functions are thus explained in a naturalistically acceptable manner. Traits are *for* their proper functions because they were selected for their performance of these functions.

My preferred formulation of the etiological theory is:

Where i is a trait of systems of type S, the/a proper function of i in S's is/was F iff a selective explanation of the current non-zero proportion of S's with i must cite F as a component of the fitness conferred by i on ancestors.

For example, it is a proper function of hearts in humans to pump blood because it is the pumping of blood by ancestral hearts which led hearts to achieve their current prevalence. It is not a proper function of hearts to fill the chest cavity, because the performance of this effect by ancestral hearts does not help to explain the current prevalence of hearts.

This formulation of the etiological theory makes it quite clear why it is possible to cite the biological proper function of a trait to explain the trait. The functional characterisation of the trait says what it is about the trait that has caused it to be selected.

The formulation as it stands does not distinguish between currently functional traits and vestiges. It captures everything which is, or has at any time been, a function of the trait. I show how to tease apart the current functions of a trait and those of which it is a vestige in the final section of this paper.

The etiological theory has been used to explain the existence of teleological properties in many other fields. Ruth Millikan has used it as the basis of a semantic theory[5]. It has even been suggested that the assignment of proper function to human artifacts is ultimately backed by a form of adaptive explanation[6]. Lindley Darden & Joseph Cain[7] noted that proper functions can be generated by any theory that involves a certain kind of selective explanation. They suggest that current theories of neural growth and of antibody production may support ascriptions of proper function. So although it was inspired by philosophical reflection on biology, the etiological theory has become more than a theory of biological function. It has become the key to distinguishing the scientifically acceptable uses of teleology from the unacceptable ones. The claim that an item exists to perform some function F must either be reduced to the claim that it has the effect F, or backed by a story in which the fact that items of this type have this effect has influenced their production or survival. Where no such story is available, no sense can be attached to the teleological claim.

In this paper, I take the etiological theory back to biology, to try and improve our understanding of the adaptive explanations that originally inspired it. The current orthodox story has it that adaptive explanation became ubiquitous in the classic evolutionary biology of the modern synthesis. Alternative forms of evolutionary explanation were backgrounded, probably because of their relative intractability. Recent attacks on adaptationism, and revived interest in the role of drift, accident and historical constraint in evolution have called into question the validity of many traditional, adaptive explanations. It is perhaps not often appreciated that these attacks on adaptive explanation also call into question the current use of the notion of biological function. If, as some authors sometimes suggest, it is naive or simplistic to explain why we have noses by sketching adaptive scenarios in which animals with better air filters or olfactory mechanisms have more offspring, then it is unclear what is meant by claiming that the function of the nose include air filtering and olfaction. If we cannot explain the nose by pointing to it's adaptive value, then the claim that the nose has these functions is on a par with the claim that its function is to support spectacles. If the teleological pretentions of biological functions are not backed by adaptive explanations, then it is unclear that they have any backing at all.

I shall suggest that a deeper understanding of proper function ascription reveals that adaptive explanations do not require naive 'adaptationist' evolutionary scenarios. A more realistic picture of evolution can still employ adaptive explanation extensively, and legitimately ascribe many biological proper functions.

3. ADAPTATION AND EXAPTATION

Gould and Vrba[8] made a well known attempt to broaden our picture of evolution by introducing the notion of 'exaptation'. They distinguished three ways in which a trait could come to have an effect E. First, it may be shaped by selection to perform E. They call this process 'adaptation'. Secondly, it may be shaped by selection to have some other effect, and coincidentally come to perform E. Gould and Vrba call this recruitment of the trait for a new purpose 'coaptation'. Thirdly, a trait may arise for non-adaptive reasons, and coincidentally come to perform E. This is also coaptation. Gould and Vrba claim that the second and third processes, those involving coaptation of an existing trait, should not be called adaptation. Instead, they call them 'exaptation'. In the first, *ad*aptive, case they say that E is a function of the trait. In the other two, *ex*aptive, cases, they say it is an effect.

I shall suggest that this is a mischaracterisation of the function/effect distinction. Once we see why this is so, it becomes apparent that the adaptation/exaptation distinction is a rather unhelpful way to taxonomise evolutionary processes. First, however, I shall consider exactly how Gould and Vrba draw the adaptation/exaptation distinction. I hope it will then be apparent what has led them astray.

Process	Character		Usage
Natural selection shapes the character for a current use - adaptation	adaptation	aptation (adaptation + exaptation)	function
A character, previously shaped by natural selection for a particular function (an adaptation), is coopted for a new use - cooptation	exaptation		effect
A character whose origin cannot be ascribed to the direct action of natural selection (a nonaptation), is coopted for a current use - cooptation			

Table 1

From Gould and Vrba, 1982. Reproduced by kind permission of the authors and publishers.

At some points in their article where Gould and Vrba say that a trait is an exaptation relative to some effect E just in case E was not the reason for the original spread of the trait in the population. This seems a perfectly clear distinction, although, as I will argue, it's not a very useful one. At another point, however, they say that things are exaptations because they are "fit for their current role, hence *aptus*, but they were not designed for it, and are therefore not *ad aptus*"[9]. It is this second way of drawing the distinction that reveals a fundamental confusion.

It is, of course, a truism that adaptation doesn't really involve design. 'Design' in evolution is a metaphor for the effects of selection, and all that happens in selection is the proportional increase or decrease of an existing trait, which can have no ultimate origin but random mutation. But despite the truistic nature of these facts, Gould and Vrba seem to want to distinguish an earlier phase, in which a trait is 'designed' or 'shaped' for something, from later phases in which it is merely found useful for other things. This causes them to overlook the critical fact that the 'exaptive' selection of a trait for a new purpose is as valid a source of selective explanation as the original 'adaptive' selection of the trait. In fact, Gould and Vrba's first category, adaptation, is just a special case of their third category, the exaptation of a trait that has arisen for non-adaptive reasons. Every trait arises by random mutation, and exists in the population at some low frequency waiting for selection to increase its frequency. I hope that this does not sound like nitpicking. If it does, bear with me while I draw out it's not inconsiderable implications.

I said above that Gould & Vrba had mischaracterisd the function/effect distinction. According to the etiological theory, an effect of a trait is a function if the fact that the trait has that effect can be used to explain the current prevalence of the trait. On this account of function, there will be functions which are merely effects on Gould and Vrba's account. The disputed cases are effects which played no role in the early selective process that initially spread the trait through the population, but which played a role in sustaining or increasing the prevalence of the trait under later ecological conditions.

Classic examples of such functions are to be found in Darwin's account of the evolution of emotion[10]. Darwin hypothesises that many expressive behaviours were initially favoured for some utilitarian purpose, and later retained because they had acquired a role in intraspecific communication. Thus, for example, he suggests that the baring of the teeth in primate anger may originally have been selected as a preparation for attack. It then acquired a secondary use in signalling aggression, and has been retained for this secondary function in, for example, humans, whilst becoming vestigial with respect to its original function (The notion of vestigiality with respect to a given function is formally defined below).

I claim that this 'secondary' function of tooth baring is a proper, biological function. I say this because a correct selective explanation of the current prevalence of the trait must mention this function. Gould and Vrba, however, are committed to calling it an effect, because it is an exaptation. What is interesting is how they get themselves into this position. They do so by failing to distinguish what we might call the two *phases* of exaptation. In the first phase, which is what they call coaptation, a trait T which has spread from it's initial mutation either accidentally or by adaptation for some other purpose, turns out to enhance an animals fitness by performing some function E. I agree with Gould and Vrba that in this phase E is an effect, and not a function of T. In the second phase, which they fail to distinguish, the enhanced fitness due to E effects the prevalence of T in the population. It is after this second phase that I claim that E is one of the proper functions of the trait. I fail to see that there is any relevant assymetry between selection finding a trait that has arisen by initial mutation and selection finding one that has been subject to drift or prior adaptation. The result in both cases is an increase in the prevalence of the trait in the population over what it would otherwise have been. In both cases, valid selective explanations of this prevalance can be stated using the selected effect. The second phase of exaptation must be regarded as a form of adaptation. If we want to distinguish later selective processes from those which fortuitously happen to have latched onto a mutation first, we might call it secondary adaptation, or, to continue the proliferation of latinate terms, 'exadaptation'.

Gould and Vrba are aware of the need for a way of representing evolutionary processes in which traits successively acquire a number of functions. But because of their failure to distinguish the two phases

1st Process	2nd Process	Character	Usage
Natural selection spreads the trait from initial mutation(s)		Adaptation	Function
A character previously spread by selections is coopted	Selection spreads the trait because of its coopted role	Exadaptation	Function
A character previously spread by another means is coopted	Selection does not spread the trait because of its coopted role	Exaptation	Effect

Table 2

of exaptation, they have to factor a trait down into a number of sub-traits, each of which has a single function. To see how this is done, it is necessary to have the right picture of an biological trait. For the purposes of evolutionary theory a notion of trait is required in which each trait has a unique evolutionary history. It follows that functional or morphological classifications of traits are unsuitable. We cannot infer from the fact that two traits do the same thing or look the same that they have a common evolutionary history. The eyes of vertebrates and of cephalopods are similar to a quite remarkable extent, yet it is generally supposed that they are the result of convergent evolution from quite different ancestral sources. The notion we require is that of a homologous trait. Two traits are homologues if they are derived from a common ancestor. A homologous trait is a character that unites a clade. It follows that traits are nested one within another in the same manner as clades (see fig.1).

A biological trait, in the sense most useful to us here, is a character that unites a clade. Every species in the clade either has the trait, or is descended from a species that has the trait.

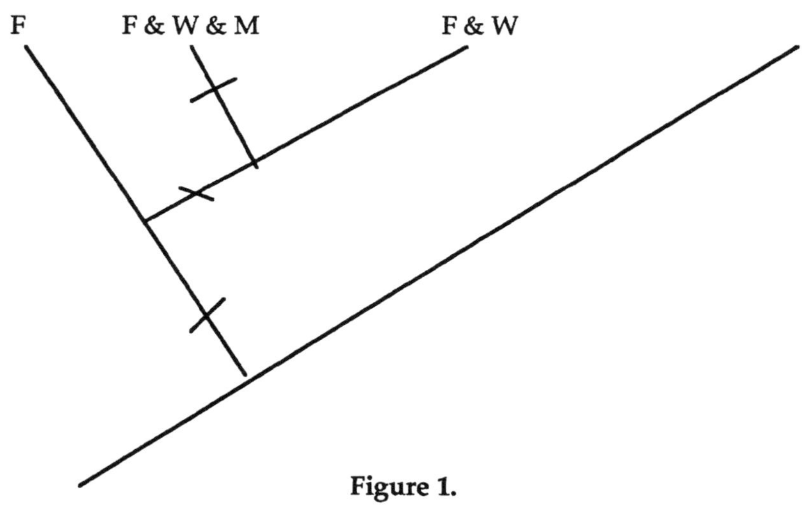

Figure 1.

Gould & Vrba apply their theory of function to a scenario for the evolution of flight. Their story can be modelled by the figure above. Proto-archaeopteryx develops feathers to catch insects. These allow primitive flight. Archaeopteryx develops wings, and because of this, its feathers are modified in order to fly better. The resulting wings and feathers allow the modern Black Heron to catch fish by shading the water (mantling). Gould & Vrba give the following representation of this story. The Heron's feathers have the *function* of catching insects, and the *effects* of allowing flight and mantling. It's wings and certain special traits of it's feathers have the *function* of allowing flight and the *effect* of allowing mantling. Neural circuits mediating mantling may have the *function* of catching fish.

The main problem with this account is the fact that it can only allow the secondary modifications of the feathers to have the function of allowing flight. Any other features of the feathers which assist flight do so only as an effect. This is because, on Gould and Vrba's story, nothing ever acquires a new function unless it is physically modified, and even then, it is only the modification that acquires the new function. This feature of their account is quite unecessary. By introducing the notion of exadaptation we can allow that feathers have functions such as flight and thermo-regulation, even if they were originally selected for catching insects. All that is required is that deficiencies in flight and thermoregulation, rather than in insect catching, have for some time been the major disadvantages preventing the spread of mutations away from standard feather forms. Whenever there is active selection there will be selective explanation, and therefore proper functions.

My preferred representation of the Heron's situation, taking Gould and Vrba's hypothetical history as given, is as follows. The Heron's feathers have thermoregulatory and aerodynamic functions, as well as being vestiges of insect catching. They probably don't have mantling functions because it is unlikely that feathers are any more prevalent in Herons than they would be if mantling were not practised. The role of feathers in mantling is therefore an effect. The role of wings in mantling is also an effect, and for the same reasons. The neural circuitry mediating mantling behaviour has the function of catching fish because this is what has caused it to achieve it's current prevalence.

The inadequacy of Gould & Vrba's account is even more clear in the case of Darwin's facial expressions discussed above. Here there is no reason to suppose that the trait is modified when it acquires its new function. Even supposing that the tooth-baring response is modified after it acquires its new function in intra-specific communication, the most likely source of change in this case would be atrophy due to loss of the original function, not modification for the new function. It would be very odd to regard the fact that the expression is somewhat reduced in vigour as an adaptation for communication, whilst refusing to regard the expression itself as anything other than a vestige of an adaptation for aggression. But on the Gould and Vrba account this atrophic modification is the only feature that can be regarded as having a communicative function.

The lesson I want to draw from this discussion is that the successive periods of selection through which a trait passes can confer a comparable succession of functions, one for each effect of the trait which has played a role in its selective history. However, this conclusion cannot be drawn purely on the basis of an analysis of the notion of proper function. The most that philosophical analysis can do is to describe a coherent, methodologically sound notion of function. Whether there are any such functions is an empirical question. In particular, the succession of functions I have described will only occur if traits remain relatively open to selection after their initial spread through the population. If a trait is maintained at fixation because of it's role in some basic physiological process, or in embryology, then changes in the utility of the trait due to the changing environment will not affect the prevalence of the trait in the population. In such a case, the trait never moves out of the first stage of exaptation - coaptation - into the second stage - exadaptation.

Despite this rider, the prospects for adaptive explanation look somewhat better than they did in the light of Gould and Vrba's original discussion. If adaptation were restricted to the initial spread of the trait in the population, it would be difficult to determine what most interesting traits are adaptations for, as this would require knowledge of the ecology of distant periods. There would be little connection between what traits are adaptations for and their current ecological roles. It is not suprising that, on this picture, 'adaptationism' is a term of derision. However, once it is recognised that proximal selective processes also yield proper functions and adaptive explanations, such explanations become more epistemically accessible. They become part of a general strategy of explanation by successive stages of selection, each historically constrained by the results of previous stages. Each stage confers functions on a trait, and leaves it's traces in the vestigial functions of the trait at later times. The main threat to this new, broader notion of adapative explanation is, as just remarked, the embedding of traits in the organisms growth and functioning in a way that insulates them from further selection. How common this phenomena is must be determined by empirical research, rather than philosophical speculation.

4. ACQUIRING FUNCTIONS AND ACQUIRING TRAITS

I tried to show in the last section that adaptive explanation has had a bad press due to the failure to recognise the importance of secondary adaptive processes. This confusion is aided by the tendency to slip into thinking that selection can somehow explain the origins of traits, whereas, in reality it can only explain their spread. In any conventional model, the 'design' of a trait is nothing more than the spread of successive sub-traits, nested within one another in the fashion illustrated above, and each arising through undirected mutation.

The same tendency to distinguish 'design' from mere selection is found in early philosophical criticisms of the etiological theory of functions. Philosophers such as Christopher Boorse[11] took an early version of the theory, due to Larry Wright[12], to propose that the functions of a trait are the effects which 'caused its evolution'. On the basis of this misinterpretation they argued that the function of a trait on the etiological theory would be whatever it was first selected for. This, of course, would not only be very difficult to discover, but even if it could be discovered, might not be the current function of the trait. The function of the penguin's wing is to assist its swimming, but this is not what he wing originally evolved for.

But as I have tried to show, this criticism is not well taken. The acquisition of traits and the acquisition of functions are quite seperate processes. A species may acquire a trait which has no function at all, perhaps due to it's genetic linkage with an advantageous trait, and this can acquire functions when the selective environment changes. It can acquire a trait with one set of functions and then add others, as the penguin has with its wings. The trait may become a vestige with respect to its original functions, leaving only the new ones, as also occurs in the case of the penguin's wing. The acquisition of functions depends upon selective explanations, but not upon selective explanations of the acquisition of traits. Whenever a trait has an effect which enhances the fitness of its bearers sufficiently to affect its representation in future generations, it acquires a new function.

In this section I consider a more recent attempt to challenge the legitimacy of adaptive explanations of the current states of organisms. Despite its greater biological sophistication, I believe that it falls into the same trap. Robert J. O'Hara suggests that many biologists have not come to terms with the historical nature of their discipline. They have not adopted certain habits of thought he labels 'tree thinking'. According to O'Hara, a question like, "Why do human beings have the appendix?" is a pseudo-question, like the primitive cosmologists question, "Why are there nine planets?" O'Hara calls these illegitimate questions 'state questions'. He suggests that they should be replaced by legitimate 'tree thinking' questions like, "Why did the branching occur at which the ancestor of these creatures acquired the appendix?" In O'Hara's words:

"To ask why certain species have a particular attribute is to suggest that that attribute is a derived character uniting them in a clade, and that the appearance of the character is the thing for which explanation is sought" (1988, p150-151)

It is easy to see that this prescription has the same effect as Gould and Vrba's refusal to allow the acquisition of function without physical modification. Functions can only be acquired at the nodes of the cladogram. They cannot be acquired as we travel along a branch. While I agree with much of what O'Hara says in his paper, I cannot agree that this is the only sensible question which can be intended when someone asks why a species has a trait. It is also legitimate to ask why the species has not lost the trait. More generally, it may be asked why the trait exists in the population at it's current level of prevalence, and not at a higher or lower level[14]. There can be illuminating answers to these questions which require knowledge of only short stretches of the organisms evolutionary history. Many of these answers will be given in the form of ascriptions of function, either current or vestigial.

A good example of a 'state question' of the kind criticised by O'Hara would be one asked, and answered, by Darwin (1872). The question is: Why do human beings bare their teeth when they are angry? The answer is twofold. First, they are members of an order many of whose members bare their teeth when they are angry. This suggests that tooth-baring was selected in a common ancestor. Plausibly, it was selected as a preparation for attack. This is the part that O'Hara would approve of. Secondly, however, we should note that tooth-baring has a communicative function in humans. This explains why the trait is maintained in the population, and does not atrophy. This twofold form of explanation is used by Darwin to explain much of our repertoire of emotional expressions. I suggest that it is a form of explanation whose applicability is very wide indeed. It may be as interesting, and as legitimate, to give selective explanations of the maintainence of a trait in a population as of its initial spread.

Once again, however, the applicability of the pattern of explanation we have described depends on certain empirical presuppositions. The second half of Darwin's explanation of an emotional expression is required because it is supposed that an adaptation which is no longer of use to the organism would not be indefinitely preserved, but would tend to decrease in prevalence, and, where retained, to atrophy (that is, to lose traits nested cladistically within the first trait, and which contribute to its functioning, or to acquire new modifications which interfere with functioning). This presupposition is another, major hostage to fortune. If it is false, then the realm of selective explanation is much curtailed. No explanation would be needed of the presence of a trait in a population other than the explanation of it's initial introduction, and considerations of heredity. Gould and Vrba's initial adaptations would be the only processes giving rise to proper functions.

The idea that unused traits atrophy is suggested by observation of such striking patterns of evolution as the loss of pigmentation and sight in a wide range of troglobytic (cave-dwelling) species. Two explanations have been offered for this phenomena (also referred to as 'rudimentation' or 'regressive evolution'). The first is simply the build up of mutations affecting the vestigial organ in the absence of any selection against them. The second, stressed by Weissmann through his concept of the 'competition of parts', is the positive selection against vestiges resulting from the drive for developmental economy. This approach suggests that since it takes energy to build an organ, mutations which reduce or eliminate useless organs will be favoured on economic grounds. However, neither of these explanations of atrophy and loss are well established. It is worth noting here the widespread view that the retention of atrophic vestiges can be explained by their role in organising embryological development. On the theory which I develop in the next section, this will result in their being vestigial relative to their original function, but acquiring, or retaining, a function in development.

The idea that unused traits tend to atrophy turns out to be vital in defining a workable notion of a vestige. In the account of vestiges offered in the next section, I will assume that rudimentation is a real phenomena. Whether this is so is, once again, not a matter to be settled by philosophical speculation.

5. VESTIGES

It is easy to point to paradigm cases of vestigiality. The ratities have vestigial wings, troglobytic animals have vestigial eyes, aquatic mammals have vestiges of their terrestrial limbs. In standard presentations, two things are said to be characteristic of vestiges. They exhibit loss or decline of function and they frequently exhibit atrophy. Both of these characteristics can be used to establish the fact that a vestige is a trait which once had a function. This fact is the foundation of my account of vestigiality.

To say that a trait has atrophied implies that it has, or had some function. It is not enough that it be smaller, or simpler, than an homologous trait in a related species. The wing of a humming bird may be smaller than the wing of some of its recent ancestors, but no-one would suppose that this is the result of atrophy! It is only when a trait is considered as having a particular function that a changes in the size or form of a trait can be thought of as atrophic.

To say that a trait has atrophied also implies that it is less capable of performing its function than the corresponding trait in previous generations of the same lineage. The imperfect wings of Archaeopteryx, or of the modern flying squirrel are not said to be vestigial, because it is not supposed that their ancestors had wings that were any more effective.

It is still more obvious that the direct claim that a trait exhibits loss, or decline of function implies that is descended from traits that had a function, and which performed it more often or more effectively.

An obvious first definition of a vestige would be that a vestige is a trait which once had a function, but which has ceased to be selected for that function for so long that it has begun to atrophy or to be eliminated from the population. Unfortunately, whilst the classic examples of vestigiality exhibit atrophy or decline, not all vestiges do so. There are two obvious cases. Firstly, and probably most importantly, a trait may cease to perform one function, but be preserved intact in virtue of it's performing a new function. As noted above, cases where a trait is retained because it has acquired a role in embryology can be represented in this way[15]. Secondly, a useless trait may subsist for an extended period of time simply because there is no genetic variation.

So a successful account must allow non-atrophied vestiges. This suggests that a vestige could be defined simply as a trait which no longer performs its function. But this is subject to other objections. First, as Neander[16] has pointed out, a *malfunctioning* trait cannot perform its proper function. But it must still have the function in order to count as malfunctioning. Secondly, it is important for traits to be able to have proper functions they cannot perform, in order to prevent functions fluctuating wildly in response to temporary environmental changes.

There is a standard reply to these problems, which appeals to the statistically normal performance of a trait in the whole population. A trait does not lose its function until it is statistically normal for the trait to be unable to perform the function. Neander[17] has pointed out that this reply fails to handle pandemic diseases, such as the viral infections of some plants. We might also note that it would not handle the widespread or universal loss of the ability to perform a function which may be induced by large environmental changes.

In the light of these difficulties, it is pleasing to discover that the idea that regressive evolution is a widespread phenomena, which I suggested in the last section, lets me avoid both sets of difficulties. If I can safely assume that useless traits tend to atrophy or decline in prevalence, then I can give an account of vestigiality which allows non-atrophied vestiges, but does not classify every trait which ceases to function as vestigial. To give this account of vestigiality, we first define the notion of an evolutionarily significant time period. An evolutionary significant time period for a trait T is a period such that, given the mutation rate at the loci controlling T, and the population size, we would expect sufficient variants for T to have occurred to allow significant regressive evolution if the trait was making no contribution to fitness. A trait is a vestige relative to some past function F if it has not contributed to fitness by performing F for an evolutionarily significant period. A trait is a vestige simpliciter if it is a vestige relative to all it's past functions. This account allows a trait to become a vestige relative to one function whilst remaining intact in virtue of it's other functions. It also allows a trait to become vestigial whilst remaining intact because of a lucky absence of mutations.

This account of vestigiality can be incorporated into the definition of proper function given above by defining a proximal selective explanation as one that involves the action of selection during the last evolutionarily significant period, or would have involved such action during that period had the mutation rate not fallen below expectation[18]:

Where i is a trait of systems of type S, the proper function of i in S's is F iff a proximal selective explanation of the current non-zero proportion of S's with i must cite F as a component in the fitness conferred by i.

This formulation captures only the current functions of a trait, and excludes those functions of which the trait is a vestige.

6. CONCLUSION

I have tried to show that adaptive explanations are not as black as they are painted. In particular, I have shown how to revise the adaptation/ exaptation distinction into a tripartite distinction between adaptation, exadaptation and exaptation. Whether there is a significant amount of exadaptation depends on whether sustaining selection is required to keep a trait from atrophying, and whether many traits are so tied up in the functioning and development of the animal that they are maintained at fixation by a small number of essential functions (such as their role in embryology) and so are not open to selection. These are empirical questions, and they ought to be left open to empirical research, not ruled out of court by a mistaken taxonomy. Both Gould & Vrba and O'Hara seem to me to have legislated in this way.

If these empirical issues are settled in a way that is reasonably friendly to adaptive explanation, then we have the following picture of evolution. Traits are selected for various effects as the environment changes. As an effect comes to have a role in selection it takes on the status of a function. As it ceases to have such a role the trait becomes vestigial with respect to that function. It is a neat picture. It would be nice if it were true.

NOTES

[1] Gould, S.J. & Vrba, E.S. (1982) 'Exaptation - A Missing Term in the Science of Form', *Paleobiology*, 8 p4 - 15.

[2] I am thinking, of course, of the dispute between V.C Wynne-Edwards (*Animal Dispersion in Relation to Social Behaviour*, Oliver & Boyd. Edinburgh. 1962) and G.C Williams (*Adaptation & Natural Selection*. Princeton University Press. 1966), who understood very well that if the group's need for population control has no influence on the evolution of a behaviour then an effect of that behaviour on population control is just that, an effect, not a function.

[3] As discussed by Robert Çummin in 'Functional Analysis', *Journal of Philosophy* **LXXII**, 1975 p741 - 765 and *The Nature of Psychological Explanation.*, Bradford/MIT Press, Cambridge, Mass, 1983.

[4] Wright, L. (1973) 'Functions', *Philosophical Review.* **82**, p139 - 168; Millikan, R. (1984) *Language, Thought & Other Biological Categories.*, Bradford/MIT Press, Cambridge, Mass. ; Neander, K. (forthcoming) 'Functions as Selected Effects', *Philosophy of Science.*; Griffiths, P.E. (forthcoming) 'Functional Analysis & Proper Functions' *British Journal for the Philosophy of Science.*

[5] Millikan, Op cit.

[6] Griffiths, Op.cit.

[7] Darden, L and Cain, J. (1989) 'Selection Type Theories', *Philosophy of Science.* **56**, p106 - 129

[8] Op.cit.

[9] Op.cit. p 6.

[10] Darwin, C. (1872)*The Expression of Emotions in Man & Animals..*

[11] Boorse, C. (1976) 'Wright on Functions', *Philosophical Review* **LXXXVI**, p70 - 86.

[12] Op.cit.

[13] O'Hara, R.J. (1988) 'Homage to Clio, or, Towards an Historical Philosophy for Evolutionary Biology', *Systematic Zoology*, **37**, p142 - 155.

[14] This seems to require that polymorphic traits be admitted when constructing cladograms. Some systematists are hostile to this, as it can be complicated. There seems, however, to be no substantial argument against the practice, and, indeed, it is hard to see how it can be avoided.

At the very least, occasional mutations in the population must be admitted, and given the frequency of polymorphisms higher frequencies have to be admitted if many characters are not to be excluded from consideration.

[15] It might also be thought that a trait which has lost its function might persist because it is genetically linked to some vital trait. This is not possible, however, because unless the trait had become more prevalent than it would be on the basis the genetic linkage alone, such a trait, however useful, will not acquire proper function. If it has become more prevalent, then when it ceases to be selected for its function, its prevalence will decline.

[16] Op. cit.

[17] Op. cit.

[18] I am grateful to Karen Neander for pointing out the necessity of this last clause. It excludes counterexamples parrallel to those which led me to include a probablistic element in my definition of a vestige. A trait might be thought to be currently contributing to fitness although it is not being selected because of an improbable absence of mutations.

Paul Griffiths,
Department of Philosophy,
Otago University.

Timothy Shanahan

SELECTION, DRIFT, AND THE AIMS OF EVOLUTIONARY THEORY

1. INTRODUCTION

According to textbook presentations of evolutionary theory, evolutionary change is a result of the interaction of a number of biological processes that together shift a population away from Hardy-Weinberg equilibrium. Among the factors typically mentioned are genetic mutation, gene flow (emigration and immigration), nonrandom mating, selection, and drift ("chance").[1] By constructing equations which factor in specific values for each of these processes, evolutionary biologists try to explain why a population follows a particular evolutionary trajectory. Hence, much of evolutionary biology is concerned with the empirical determination of values for each process, and the ways in which the various processes can and do interact with one another to produce evolutionary change.

A common presupposition of this approach is that each of the evolutionary factors just identified represents a distinct biological process (or force), clearly delineated from the rest. For example, selection is understood as a quasi-deterministic sampling process operating on the basis of heritable variation in fitness among individuals at some level of the biological hierarchy. Drift, on the other hand, is understood as the consequence of random changes in gene frequencies, usually over several generations, due to "sampling error". As such it is an inherently stochastic process. "Selection" and "drift" are concepts designed to capture two distinct but interacting biological processes.

My aim in this paper is to critically examine this standard account, focussing on the processes of selection and drift. Are selection and drift really distinct processes in nature? What purpose(s) (if any)

P. Griffiths (ed.), Trees of Life, 133–161.

does the selection/drift distinction serve in evolutionary biology? My plan is to explore these questions by isolating the critical issues involved. My specific aim is to understand the precise relationship between selection and drift both in evolution and in evolutionary theory, in the hope that doing so will shed light on the nature of this theory, including its internal structure and how it relates to the living world it is designed to render comprehensible. I will argue that although selection and drift are not as distinct as a superficial reading of the textbook account would suggest, neither are they biologically identical. Rather, "selection" and "drift" are idealised concepts which represent endpoints on a continuum of biological sampling processes. Yet, despite the lack of any clear-cut distinction between selection and drift in nature, there is good reason to distinguish them in evolutionary theory because doing so serves certain explanatory and predictive purposes. I then draw some implications of this conclusion for understanding the nature and aims (and limitations) of contemporary evolutionary biology.[2]

2. THE CONVENTIONAL DISTINCTION

As John Beatty points out, both selection and drift can be understood as sampling processes.[3] Both processes involve the differential survival and/or reproduction of biological entities. What distinguishes these processes is whether the sampling process is "discriminating" or not. Sampling is "discriminating" when the survival and/or relative offspring contributions to the next generation are affected by physical differences among individuals. To take the classic example, consider light and dark moths living in a forest which, because of nearby industrial pollution, is becoming increasingly soot-covered. In the presence of avian predators which hunt by visual recognition of prey, the dark moths will have a survival advantage in this environment. The avian predators will "sample" the moth population, discriminating among moths on the basis of a physical difference between them. Over time, the proportions of dark to light moths will shift toward the dark morph. This is of course a classic example of natural selection producing evolutionary change.[4]

Sampling is "indiscriminate" when physical differences between individuals affect neither their own survival nor their relative offspring contributions to the next generation. Beatty gives the example of a forest fire which sweeps through an area killing some organisms and sparing others without regard to physical differences between them. He writes that, "Such sampling is indiscriminate in the same sense in which the usual model of blind drawing of beads from an urn is indiscriminate — that is, any physical differences (e.g., colour) between the entities in question are irrelevant to whether or not they are sampled."[5] Repeated occurrence of such indiscriminate sampling in a finite population can result in the random drifting of properties (or gene frequencies) over several generations. Hence the name "random genetic drift".

The point of departure for my critical examination of the standard "textbook" account (as clarified by Beatty) is a modification of an example originally due to Michael Scriven but alluded to and elaborated on by many subsequent writers.[6] Suppose that two "identical" (i.e., monozygotic) twins are hiking along a mountain ridge when suddenly an electrical storm arises. ZAP! — One of the twins is struck dead by a bolt of lightning. The other twin, slightly singed, dazed but still alive, goes on to live a long and fruitful life, eventually fathering some dozen children, all of whom are equally prolific. What are we to say about this case? Is it an example of selection or of drift ("chance")? How can we decide which it is?

Most of the biologists I have posed this question to have concluded, after a moment's reflection, that it was a purely chance occurrence, and hence not selection.[7] This interpretation could be supported by the following considerations. There were, *ex hypothesi*, no genetic differences between the twins — they were genetically identical. Genetic identity does not guarantee phenotypic identity, of course, but we can assume for the sake of the example that neither twin has a developmentally or environmentally induced characteristic that would make it more or less vulnerable to a lightning strike. For example, neither twin was given pituitary growth hormone as a child, so that it grew twice as tall as its sibling, thus making it into a kind of natural lightning rod. In the absence of such phenotypic differences between the twins, the only recourse is to attribute the fateful occurrence to drift (chance).

My strategy in the next section will be, first, to defuse the most common argument that asserts that this cannot be a selection event by challenging the presupposition it rests upon, and then, second, to give a positive argument for the claim that the twin case is a selection event. This strategy will serve to bring to the surface the issues at stake in distinguishing selection and drift.[8] What is at stake here is not simply the best interpretation of a purely hypothetical abstract example. How one interprets the twin case bears on the validity of the selection/drift distinction in general. The case of the twins merely allows us to focus our attention on the relevant issues. Insights gained from this example can then be applied elsewhere in understanding evolutionary biology.

3. SELECTION OR DRIFT?

One might argue that the case of the travelling twins cannot be a selection event, by reasoning as follows. Both selection and drift are sampling processes. What differentiates these two processes is that the former discriminates on the basis of *fitness* differences among biological entities, whereas the latter does not. Selection is a form of discriminate sampling on the basis of fitness differences among biological entities, whereas drift is a form of indiscriminate sampling (i.e., not on the basis of fitness differences) among biological entities. So, according to Beatty[9], "the two physically identical twins, who must be physically disposed to contribute the same number of offspring, are equally fit. Hence we can say of the lightning that killed one and spared the other, that it did not sample the twins on the basis of their fitnesses —it was not an agent of natural selection. It was clearly an indiscriminate sampling agent — indiscriminate, that is, with regard to physical differences between the organisms sampled." This argument rests on several important claims, all of which need to be critically examined, *viz.*: (1) fitness differences are essential for selection, (2) the twins do not differ in fitness, and (3) there are no physical differences characterizing the twins with respect to which the sampling could or did discriminate between them.

Let us begin with the second claim. That there are no fitness differences in the twin case can be shown in two ways, providing the basis for two arguments that this was not an example of selection, but rather one of drift: (1) *Supervenience Argument:* Fitness is a *supervenient* property.[10] Hence, if the underlying properties of two biological entities are identical, they also have identical fitnesses. In the case of the twins, their underlying properties were identical. Hence they did not differ in fitness. Selection requires fitness differences. Hence it was *not* an example of selection. It was either selection or drift. Therefore, it was drift. (2) *Propensity Argument:* Fitness is a *propensity* (or disposition) to survive and produce viable offspring.[11] *Subargument #1:* Identity of underlying properties entails identical propensities. In the case of the twins, their underlying properties were identical. Hence, their propensities for survival and reproduction were identical. Hence, they did not differ in fitness. Selection requires fitness differences. Hence it was not an example of selection. Therefore, it was a case of drift. *Subargument #2:* Propensities are *ascribed* to entities in virtue of observed correlations between the possession of certain properties and certain effects (e.g., "fragility" and breaking easily when subjected to a sudden force). In the case of the twins there was no basis for ascribing different propensities for survival and/or reproduction to the twins because there were no properties of the twins that would allow us to determine *before the event* which one would be more likely to die from a lightning strike. One could argue that because the case of the twins is an utterly unique event, there is no basis for ascribing different propensities to them. Hence, there was no basis for ascribing fitness differences to them. Selection requires fitness differences. Hence it was not an example of selection. Therefore, it was a case of drift.

My response to these arguments has several parts. First, I agree that there are no fitness differences between the twins. Fitness is best construed as a supervenient propensity for survival and/or reproduction. The twins did not differ in fitness because there was no basis, before the lightning strike, for predicting their different fates in the context of an electrical storm.[12] Claim (2) above is correct.

My (first) disagreement concerns the claim that fitness differences are necessary for (and not just conducive to) selection events. I want to claim that contrary to many presentations of evolutionary theory, fitness differences are *not* necessary for selection (and hence cannot be used to distinguish selection and drift). My claim builds upon, and I think follows from, the very fact that fitness is a supervenient propensity for survival and/or reproduction.

As a *supervenient* property of an organism, fitness is grounded in an indefinite number of underlying properties that affect that organism's ability to survive and/or reproduce, properties such as disease resistance, visual acuity, protective colouration, fleetness, position in a dominance hierarchy, etc. In any given interaction with the environment, only a small subset of the properties determining an organism's fitness *actually* contribute to an organism's survival and/or reproduction. Most play no role whatsoever. When being pursued by a predator, an organism's coordination, reflexes, and speed become relevant factors, but the average litter size it is prone to produce may not. "Fitness" is a gauge of how biologically successful an organism is *likely* to be given its particular suite of characteristics and the environments it is likely to encounter. As such, it does not directly determine an organism's fate. What *directly* determines an organism's fate is the set of *actual* environmental challenges it encounters and its particular resources for dealing with these challenges.[13]

A consequence of the claim that fitness is a supervenient property of an organism is that organisms having identical fitnesses may nonetheless exhibit quite different underlying properties — differences which may determine differences in biological success. Fitness, as such, is not a property of an organism which directly interacts with the environment. As Sober (1984) has cogently argued, fitness is causally inert.[14] Yet selection is a causal process, one involving interaction between some aspect of an organism's phenotype and some "critical factor" in the environment.[15] It follows that selection can occur in the absence of fitness differences. Not only do fitness differences not guarantee selection, but fitness differences are not even necessary for selection.[16]

As a *propensity*, fitness describes an entity's typical behaviour when placed in specified circumstances. Propensities can be ascribed to entities in either (or both) of two ways. One can place a given entity in the desired circumstances a number of times, observing what effect this has, or one can note what properties an entity exemplifies, and observe how similar entities behave when placed in *similar* circumstances. The first method is more direct, but only works with entities which survive the treatment intact. The second method is more indirect, and only works when one can be sure that both the entities and circumstances considered *are* sufficiently similar to warrant an inference from one to the other. In either case, propensities are ascribed to entities in virtue of observed correlations between certain properties and specific effects.

A consequence of interpreting fitness as a propensity, and claiming that fitness differences are necessary for selection, is that no *unique* sampling event can be identified as a selection event. On this view, to identify a selection event requires identifying fitness and hence propensity differences. But propensities can only be ascribed in either of the two methods just outlined. In a unique sampling event, neither method is available. If the entities in question have never been placed in the specified circumstances before, and if the uniqueness of the situation precludes analogy with other "similar" cases, then there is no basis for ascribing differences in propensity, and hence in fitness, to the entities. But if selection requires fitness differences, then there are no unique selection events. On this view (which I think is mistaken), all selection phenomena are by definition repeatable. Singular selection events are a contradiction, and hence impossible.

Hidden in the propensity argument is the assumption that the correct description/interpretation of a sampling event is determined by locating the event in a larger pattern of "similar" events in which it is thought to be an instance. By definition, a selection event is part of a regular pattern, and one instance cannot constitute a pattern. Hence, a series of selection events is composed of individual events, no one of which is, taken individually, a selection event. It would commit us to saying that any given sampling event in nature cannot be, and hence cannot be identified as, a selection event *until* it can be seen in the context of many other "similar" sampling events.

If natural processes are to be understood as objectively real occurrences, and selection is a natural process, then selection events occur in nature quite independently of their description by biologists. Although their identification is made considerably *easier* when placed in the context of similar events of which they can be seen to be an instance, this does not entail that selection events do not occur *except* as parts of identifiable patterns. If a sampling event is unique in its characteristics, it may nonetheless be a selection event, despite the fact that it does not occur as part of an identifiable *pattern*, and hence does not occur because of an identifiable *propensity*, and hence does not occur because of identifiable fitness differences. Again, fitness differences are not necessary for selection events.

Finally, recall that there were three central claims in Beatty's argument that the twin case is an example of drift rather than selection, *viz.*, that: (1) fitness differences are essential for selection, (2) the twins do not differ in fitness, and (3) there are no physical differences characterizing the twins with respect to which the lightning could or did discriminate between them. I accepted the claim that the twins do not differ in fitness, but have been arguing that the claim that fitness differences are essential for selection is mistaken. I turn now to the third claim. Are there any physical differences characterizing the twins with respect to which the lightning could or did discriminate between them?

Discussing the twin example, Mills and Beatty (1979) write that, "Surely in this case there is no difference between the two organisms which accounts for their difference in reproductive success."[17] Hence they conclude that this was not a selection event. I want to claim that, on the contrary, there *is* a physical difference between the twins that accounts for their differential biological success. Contrary to initial appearances, there is *one* respect in which the twins differ from one another — location on the mountain. It was the fact that they occupied different spatial coordinates when the lightning struck that was the crucial factor that resulted in their differential survival. The twins differed in *locational properties*, resulting in differential survival, and hence in selection.

It might be objected that being at a particular place at a particular time is not a property of an organism. Surely "locational properties" are pure fiction. But consider that location, like size, is a *spatial* property of an entity. Size is a function of volume and configuration, while location is a function of relationships to other entities in some coordinate system. Spatial properties such as size are recognised as important factors in evolution.[18] Likewise, biological success is *almost* always a function of where an individual is relative to other entities in its environment. Where an organism finds itself can be decisive for its moment to moment prospects for survival. To take but one example, a common explanation for schooling in fish is that predators generally take a disproportionate number of individuals from the periphery of the school. This results in a selective pressure to move toward the interior of the group, hence the tightness of the group and apparently coordinated group maneuvers. Such examples are not unusual. In an otherwise genetically and phenotypically homogeneous population, locational properties can be the decisive factor determining an individual's survival. Selection can operate on locational properties as easily as on any other kind.

It might be objected that locational properties, as I have described them, cannot play the role that I assign to them in selection events, because they do not constitute an *unchanging* part of an organism's phenotype. That is, one might argue that to ground a selection event, the properties in question must be stable, standing properties of an organism which persist from one environment to another. So, in the example of the travelling twins, had it been the case that one of the twins (the one that got zapped) had a stable *disposition* to walk along a more exposed (and hence more vulnerable) path during the electrical storm, then we could ascribe the differential survival to selection. In the example as described, however, there is no such disposition. That one twin got zapped while the other escaped unharmed seems more a matter of luck or coincidence than of selection. Hence it is not a case of selection.

To see what is mistaken in this argument, consider the following hypothetical experimental set-up. Imagine an artificial selection regime consisting of two containers of insects and a disk which spins freely. I put the containers on opposite sides of the disk which is then spun around. I stop the disk, remove the container from the left hand side, and kill all of its inhabitants. I then breed the insects in the remaining jar, form a new group that is then placed in the empty jar, spin the disk again, and again kill all the insects who end up in the left hand container. This procedure is repeated fifty times. Question: Is this selection? On what grounds could it be denied that it is?

In the artificial "selection" experiment above, organisms (or groups) were indeed selected on the basis of a locational property —

being on my left. The fact that a member of *Homo sapiens* was the selective agent is irrelevant to the analysis of the process. Those insects just happened to find themselves in an unusual environment. Plants on the edge of a volcano, or fruit flies in a geneticist's lab, also find themselves in unusual environments. It is true that in the case described one would not be able to *predict*, in advance, which organisms would survive and which would not. But this has more to do with our ability to look ahead than it does with the nature of biological events. One might argue that the property of being on my left is not a real property, because it is constantly subject to changing environmental factors. Besides begging the question, this objection overlooks the fact that any property, including all dispositional/intrinsic properties, can only be such in relation to environmental/external conditions. Being light-coloured is impossible without the presence of light.

I conclude that the organisms on the rotating disk do differ with respect to a property (the property of being on my left at time *t*), but they do not differ with respect to any "inhering property," such as colour, weight, visual acuity, etc. They don't differ with regard to *fitness*. What follows from this observation is simply that prediction is much more difficult in such cases, not that selection does not occur. Returning now to our hypothetical twins, we can say that they differed in what proved to be a very important property — location on the mountain. Because of this property difference, one was killed and the other survived. The fact that the property in question is not an enduring, stable characteristic of the organism is quite true but irrelevant. Hence this is, contrary to the common view, a selection event.[19]

Even if it is granted that locational properties are real and possibly important in some contexts, one could still argue that neither the case of the twins nor that of the insects on the rotating disk were examples of selection, because in neither case were the locational properties in question *heritable*. The surviving twin will not pass on to his offspring the property of being at location *x* on the mountain, nor the even more useful property of being in a safe location during an electrical storm. Likewise, the insects ending up in the right-hand container do not pass on to their offspring the property of ending up in the right-hand container — after all, *some* of the offspring of the "lucky" insects will, because of the experimental set-up, be placed in a container that ends up on the left-hand side during their "turn at the wheel"! Evolution by natural selection requires that the properties being selected be heritable. Hence neither the twin case nor that of the insects on the rotating disk is a selection event.

This argument rests on the mistaken assumption that selection only operates on heritable properties. It fails because it confuses the requirements for *evolution* with those for *selection*. Whereas evolution by natural selection does require heritable characteristics, selection (more precisely, phenotypic selection) does not. The issue of heritability is irrelevant to identifying a sampling event as a case of selection.[20]

Shifting attention from the nature of the properties that are operative in selection to the nature of selection itself, one could argue that the twin example is not a case of selection, as follows. Selection is an interaction between the properties of biological entities and the environment. To say that *location* is a property of an organism is to erase the organism-environment distinction, because it makes the environment *part* of the organism. But if there is no organism-environment distinction, then there is no room left for *interaction* between the biological entities and the environment. At most one could have interaction between some properties of an entity and some *other* properties of that entity. Hence, so long as selection is understood as an organism-environment interaction, the twin case is not an example of a selection event.[21]

This argument misunderstands the claim that location is a biologically significant property of an organism. Being at location *x* at time *t* is a property of an organism. Because of this property an organism may be affected by some factor in its environment — e.g., a predator or other natural agent. But to say that location is a biologically significant property of an organism is not to claim that *all* aspects of an organism's environment are part of its phenotype, nor that every environmentally-based property of an organism is biologically or selectively significant. The lightning which interacted with the phenotypes of the twins was a part of the twins' environment, not a part of their phenotypes. It was in virtue of this separateness that an interaction between it and the twins' phenotypes was possible. Some environmentally determined properties of the twins with respect to which they differed (e.g., in the shade or in the light, facing northeast or southwest, etc.) were not, in this case at least, biologically (i.e., selectively) significant. Just as components of fitness must be distinguished in order to understand why a particular selection event occurred, so too must *components* of an organism's phenotype (e.g., its location) and its environment (e.g., a lightning bolt) be distinguished. Hence the interaction objection fails.

A related objection would argue that it is true that there must be *some* difference between the twins that accounts for their differential survival, but hold that it is not a difference in their *properties*. The twins do not differ in so-called "locational properties," or in any other properties for that matter. Rather, their differential survival is to be explained by recognizing that they simply occupy different *environments*. Twin *A*'s properties in environment *X* resulted in his survival, whereas twin *B*'s (identical) properties in environment *Y* resulted in his demise. This is no different from saying that a fast gazelle in one environment (say one lacking cheetahs) survives even though an equally fast gazelle in another environment (one populated by cheetahs) is predated. After all, it is obvious that whether or not a property of a biological entity makes any selective difference depends on the environment in which it occurs. In the case of the twins, they simply inhabit different environments, and properties that were an asset in one environment were a liability in another. In general, for two biological entities to participate in the same selection process they must be subject to a common environment. If it is claimed that the twins have different "locational properties," then this should be reinterpreted to mean that they occupy different environments. But if so, then they cannot be participants in the same selection process. Hence the case of the twins is not an example of selection.

There are two responses to this argument. First, to say that the twins differ in locational properties does not entail that they occupy different environments. If the relevant environment is (arbitrarily) *defined* as the mountain upon which they are hiking, then they occupy the same environment, and the difference in their fates must be explained on some other basis, e.g., on the basis of property differences between them.[22] Secondly, this objection, if correct, would prove too much. No two organisms inhabit identical (micro-) environments. Any two organisms will inhabit environments in which many, but not all, parameters are for all practical purposes the same. If it is argued that for selection to occur organisms must inhabit the identical environment, then selection never occurs. This objection either demonstrates too little or proves too much. Either way, it fails to establish its claim that the twin case is not an example of selection.

Finally, one could object that if the twin case is really an example of selection and not drift, then evolutionary theory falls prey to the charge of "panselectionism". If the twin case is really an example of selection, then it looks like all differential survival and/or reproduction are cases of selection. We can know, *a priori*, that a case of differential survival and/or reproduction is a case of selection without even examining the details. The result is that evolutionary theory is emptied of all of its empirical content. It becomes a trivial game of words, a pseudo-science on a par with astrology, palmistry, and alchemy.

I hope that is it obvious how silly this objection is. Nowhere am I claiming (or even presupposing) that (i) selection is all there is to evolutionary theory, nor (ii) that all evolutionary change is due to selection, nor (iii) that selection is even the most important factor in evolution. First, as stated at the outset of this paper, evolutionary theory attempts to factor together a number of distinct biological processes to account for evolutionary change. Among these factors are genetic mutation, gene flow, nonrandom mating, selection, and drift. My concern here is with selection and drift. I have said nothing about the distinctness, identity, or importance of the other factors.

Secondly, what I *am* claiming is that selection and drift, while conventionally distinguished as two distinct but interacting evolutionary processes, cannot in fact be distinguished so cleanly. Nowhere in this claim is it presupposed that selection is the only or even the most important evolutionary process, much less that selection is all there is to evolutionary theory. Clearly it is not.

Beatty is right to distinguish selection from drift by saying the former is discriminate sampling whereas the latter is indiscriminate sampling. But what is being sampled are *properties*. A sampling process may be undiscriminating with regard to some properties, but discriminating with regard to others. In the case of the twins, there is selection with regard to a spatial property with respect to which they differ (location relative to the lightning bolt), but not with regard to many others with regard to which they also differ (location with regard to Greenland, Margaret Thatcher, and the Hollywood Bowl). My claim is that in any case of differential survival and/or reproduction, there will be some property difference with respect to which selection is operative. Thus all supposed cases of drift can be reinterpreted as cases of selection. But it does not follow from this that we can know ahead of time which property was operative in any given sampling event. There is still plenty of room on the view advanced here for the empirical investigation of selection events to determine (i) *which* properties were actually operative, (ii) *how* they contributed to differential biological success, and (iii) what *effect* this event might have on questions of interest to evolutionary biologists. Knowing that all sampling events are selection events, far from putting evolutionary biologists out of work, instead makes it clear how much work there is still to do in explaining biological phenomena.

4. IMPLICATIONS

It is time now to summarise the argument thus far, and to indicate some of its implications. Conventional textbook presentations of evolutionary theory represent selection and drift as two distinct but interacting evolutionary forces (or processes). They are to be distinguished in terms of whether or not the sampling is caused by fitness differences. I argued that this method cannot be used to distinguish selection and drift, because *neither* process is caused by fitness differences. The twin case was introduced as a paradigmatic example of drift. Although commonly interpreted as a purely "chance" event, I argued that it is better interpreted as a case of selection. But if so, then the selection/drift distinction is called into question. Perhaps selection and drift are not biologically distinct processes.[23] Perhaps they are just two different descriptions for the *same* process. Just as "Morning Star" and "Evening Star" *appear* to describe two different entities, but are in fact identical (both are Venus), so too with selection and drift. If the case of the twins (which appears to be an example of drift if anything is) is in reality an example of selection, then it seems that the selection/drift distinction that underlies most textbook presentations of evolutionary theory does not capture a real distinction in nature.[24] The question is, are selection and drift merely identical processes with different names?

I think that an unqualified affirmative answer to this question would be near to the truth, but misleading. It may be true that selection events and drift events are indistinguishable when viewed as isolated, individual events, but it may also be true that they can be distinguished with the benefit of hindsight. We are often interested in the evolution of *adaptations*. Certain kinds of natural events lead to adaptations, whereas others do not. For the evolution of adaptations, repeated sampling is necessary *in the same direction*. That is, there has to be a consistent correlation between possessing certain properties and survival/reproduction. The property in question also has to be heritable. A property which sometimes aids survival and sometimes not, or is not heritable, will not evolve into an adaptation.[25] We call those events which do lead to adaptations "selection" events, and those which do not we ascribe to "chance" or "drift". (This is even true in Sewall Wright's shifting balance theory, for which drift without selection does not produce adaptations.) To be more precise, selection events are those sampling events that are judged to have a fairly high potential for producing adaptations, whereas drift events are those sampling events that are judged to have a fairly low potential for producing adaptations. The difference is one not of kind, but of *degree*. Every kind of sampling event is "unique" the first time it occurs. But if repeated for a sufficient number of times over enough generations, it *becomes* an example of selection which may result in the evolution of adaptations, or, alternatively, it leads nowhere, adaptively speaking, and hence is described as drift.[26]

This approach suggests an alternative to the two theses considered so far. According to what we can call (for lack of a better label) the *Distinctness Thesis*, selection and drift (along with assortative mating, mutation, and gene flow) are biologically distinct processes. Any given biological sampling event is either selection or drift, but not both. Selection and drift are distinguished in terms of whether sampling involves differences in fitness among the biological entities being sampled:

[Drift] <————interaction————> [Selection]
(No Fitness Differences) (Fitness Differences)

This thesis was criticised via the twin example. Selection and drift are not entirely distinct, at least when viewed as individual events. Neither process requires fitness differences. The distinctness thesis is too simple.

At the other extreme is what we could call the *Identity Thesis*. On this view, selection and drift are biologically identical. All biological sampling is selection:

[Drift = Selection]
(Both are sampling on the basis of property differences.)

This thesis was defended above, but then qualified. All sampling events, including those described as drift, can be interpreted as selection events when looked at in some detail. However, some sampling events tend toward adaptive traits (selection events), while others (drift events) do not. But if so, then there is a sense in which the processes are distinct, at least with the benefit of hindsight. The Identity Thesis is also too simple.

In contrast with the two theses just rejected which treat selection and drift as either entirely distinct or entirely identical, I want to propose one which attempts to tread a middle ground. According to what I will call the *Continuum Thesis*, selection and drift are biologically identical at the level of individual (isolated) events (the point of the twin example), but distinguishable at the level of series of such events (the relevance of adaptedness), and (in their pure forms) lie on the ends of a continuum of sampling processes having different selection coefficients:

[Drift .... Selection]
(Low Selection Coefficient) (High Selection Coefficient)
(Random *re:* adaptedness) (Tending toward adaptedness)

On this view, selection and drift are distinguished only with hindsight, in terms of the kinds of effects produced. The selection/drift distinction is ultimately a *pragmatic* one, based upon what we find interesting and important and worth accounting for (i.e., adaptedness). Were visiting extraterrestrial biologists to survey the same natural events, but be profoundly uninterested in adaptiveness, they would be disinclined to distinguish selection and drift. That selection and drift *are* thought worth distinguishing by biologists suggests implications for understanding the aims of evolutionary theory.

If, as I am suggesting, selection and drift are not clearly distinguishable in nature, why then are they treated as distinct processes (or forces) in evolutionary theory? What is gained by affirming in theory what is nonexistent in nature? The answer has to do with the very nature of the scientific enterprise. Scientists, as a rule, are interested in general truths about nature (e.g., natural laws), rather than in the explanation of particular, often unrepeatable, events. The distinction between explaining the general and explaining the particular is often the basis for distinguishing science from history, and biology from natural history. It is also the basis for distinguishing selection and drift in evolutionary theory.

Notice how selection and drift are usually distinguished — on the basis of whether the sampling is caused by fitness differences (selection) or not (drift). Recall that fitness interpreted as a propensity describes the characteristic behaviour of an entity or of the properties it exemplifies. "Selection" is a term used to describe sampling processes on the basis of properties which are exemplified in a number of different biological entities, which ground processes exhibiting a repeatable pattern, the causes of which are deemed worth investigating. "Drift" (or chance) is a term used to describe sampling processes on the basis of properties that are not widely exemplified, which ground unique events, the particular causes of which are at present unknown, and are deemed too particularistic (i.e., apparently lacking in general significance) to merit serious scientific investigation.[27]

"Drift" ("chance") in evolutionary biology is a term which expresses our inability to encompass in theory the complexity of causal interactions that occur in nature. When certain properties had by a number of biological entities seem to be consistently connected with differential biological success, we say that there is selection for such properties. When there is differential biological success among biological entities which is not obviously connected with certain common properties, then we ascribe this phenomenon to "drift" ("chance"). This is especially true when dealing with small numbers of biological entities, in which case there is less room for generalizing about the causal importance of common properties (because there are less entities which instantiate such properties).

The limit is reached when there are just two biological entities. Hence the difficulty of the twin case. Here, there are no general trends discernible that would allow us to easily identify selectively relevant properties and forces. This accounts for the tendency to identify this as a drift event. But strictly speaking all selection events are similarly individualistic in nature. The results of such individual selection events are "summed up" to give us a theoretically useful (i.e., explanatory, predictive) account of the kinds of forces at work in a population. "Selection" events are just sampling events grounded in commonly exemplified properties. "Drift" events are just sampling events grounded in properties lacking generally significant evolutionary effects. Selection and drift are two different descriptions for property discriminating sampling processes. In evolutionary theory, "selection" (like "fitness") is a term used to capture generalities, to isolate the features different sampling processes have in common — thereby providing the foundation for explaining general phenomena.

The thesis I am defending — that selection and drift are not entirely distinct in *nature*, yet are clearly distinguished in evolutionary *theory* — has implications for understanding the methods and aims of evolutionary biology, and its relationship to the nature it seeks to comprehend. Evolutionary theory, like all scientific theories, faces a dilemma of competing values. Scientific theories are judged (among other things) on the basis of theoretical generality and empirical specificity. Both are required of any scientific theory, but both are not attainable to the highest degree. To generalise is to look for recognizable patterns in the often chaotic flux of phenomena. To do so requires abstracting from the particulars of specific cases. Empirical specificity, on the other hand, is concerned with the precision of fit of theory to actual, highly specific, situations. The most empirically accurate theory would be one which simply describes, in infinite detail, each event occurring in its intended domain. But to do so would hardly resemble science as we know it at all. Between the competing values of generality and specificity, the former usually takes priority. But the latter value can be just as important, depending on the purposes at hand.[28] Here, as elsewhere, values determine practice.[29]

5. CONCLUDING REMARKS

Evolutionary theory succeeds in explaining biological phenomena of general interest (adaptedness, diversity, etc.) because it utilises "aggregative" terms like "fitness," "selection," and "drift". Individual biological phenomena are explained in terms of such general concepts, or are left unexplained until they are seen to be an instance of a more general phenomenon. As Sober points out, "In a sense, individual organisms are not really what the theory [of evolution] is *about*." [30] It is about populations. "The point is not to isolate the unique constellation of factors that most precisely circumscribes the fate of a single organism. Rather, the idea is to bring out patterns that apply both within and among populations."[31] Yet nature consists of both general trends *and* such individuals. Evolutionary theory (at present) deals only with the former. Hence, there are phenomena which escape the net cast by the theory.[32] Whether such phenomena are considered important depends of course on the reference frame one selects for assigning measures of importance.

To illustrate this last point, consider two cases which seem to lie at opposite ends of the spectrum of evolutionary importance. Each case can be seen as a variation of the twin case discussed in some detail above. Consider first the commonplace observation of the mangled body of some small animal in the middle of or alongside a rural thoroughfare. Where I grew up — Upstate New York — it was common to see the remains of a opossum that was obviously hit by an automobile as it tried to make its way across the road. ("Playing dead" before the threat of a Buick is a highly mal-adaptive behaviour.) Suppose two opossums are crossing the road. One is hit by an automobile and is killed, while the other, having started the trek across the road a moment later, makes it to the other side unscathed. What are we to say about this case: Is it a case of selection or of drift (chance)? I can imagine the response that it really does not matter what we call it, because it is such an isolated, insignificant event. But viewed from another perspective, it is neither isolated nor necessarily insignificant. If I added up all the dead opossums I've seen along roadways, the total number would be impressive. How can one be sure that there is no significant selection going on here? In addition, one of the beauties of evolutionary theory is that it offers a comprehensive framework for understanding the often chaotic events in the living world. In principle, *every* event involving living beings, and especially those involving differential survival, should be describable within a broadly evolutionary framework. Viewed from this perspective, to ask whether the road-kill one is now contemplating is a case of selection is simply to apply the evolutionary perspective in a comprehensive manner. Evolutionary theory *should* have something to say about it.

Finally, consider a more "serious" example. In a recent book, Stephen Jay Gould discusses the ancient decimations evident in the Burgess Shale of British Columbia.[33] Analysis of the data suggests that certain body-plans were eliminated from the history of life, not because of any evident structural inferiority, but simply by being at the wrong place at the wrong time, i.e., by chance. A scientific explanation for the present distribution of organismic body plans will be concerned with functional morphology, laws of physiology, and general evolutionary principles. But an equally important part of the explanation will be concerned with particular episodes in the history of life which decisively eliminated certain body plans from the competition, making radiation from the others possible. Start the tape of life over again, Gould suggests, and we could easily be spectators to a totally different history of life on earth. Or, rather, we would not be spectators, because the body plan that led to chordates (and hence to us) might well have been one of the unlucky victims of blind fate. Or was it selection? When the isolated, unique event in question is something as momentous for the history of life as the decimations Gould describes, the question of whether it was due to selection or to drift, and of the relationship between them, displays its true significance.

NOTES

Acknowledgements: I would like to express my sincere gratitude to Elliott Sober, Paul Griffiths, Susan Oyama, John Endler, Linda Zagzebski, and an anonymous referee for helpful suggestions on an earlier draft of this paper.

[1] See, for example, D. Futuyma (1986), *Evolutionary Biology*, 2nd ed. (Sunderland, MA: Sinauer), pp. 85-87.

[2] As John Endler has pointed out to me, most biologists do not think of selection and drift as distinct processes, but rather as ends of a continuum. This can be expressed quantitatively by noting that selection may take on any value from zero to a large value. "We speak of selection and drift as ends of a continuum, to focus our attention on the mechanisms and to show that they operate" (Endler, personal communication). Part of my aim in this paper is to present novel reasons in support of this view, and to draw out its implications for understanding evolutionary theory.

[3] J. Beatty (1984), 'Chance and Natural Selection', *Philosophy of Science* **51**:183-211.

[4] H. Kettlewell (1973), *The Evolution of Melanism* (Oxford: Oxford University Press).

[5] Beatty (1984) [note 3], p. 189.

[6] M. Scriven (1959), 'Explanation and Prediction in Evolutionary Theory', *Science* **130**:477-482; R. Brandon (1978), 'Adaptation and Evolutionary Theory,' *Studies in History and Philosophy of Science* **9**:181-206; S. Mills and J. Beatty (1979), 'The Propensity Interpretation of Fitness', *Philosophy of Science* **46**:263-286; J. Beatty (1984) [note 3]; E. Sober (1984), *The Nature of Selection* (Cambridge, MA: MIT/Bradford Press).

[7] In this paper, I am treating drift as the "chance" factor in evolutionary theory. This is in keeping with most formal and informal presentations of the theory. The concept of "chance" has had and continues to have other meanings in evolutionary biology, such as coincidence, ignorance of causes and as "accident". I have explored these different notions elsewhere; space prohibits me from doing so here.

[8] The *class of* issues at stake in considering the selection/drift distinction all concern cases in which factors other than selection coefficients (assignments of fitness values) affect the fates of organisms, populations, and species. According to Mayr's "founder principle," for example, a new species may originate from the genetic divergence of a peripheral isolate of a population. Which individuals form the isolate, and which survive the initial displacement, may have very little to do with their selection coefficients, but may have very great evolutionary consequences in the long run.

[9] Beatty (1984) [note 3], p. 192. A sampling event may be "indiscriminate" with regard to some properties, but not with regard to certain others. Balls drawn out of an urn by a blindfolded man have been sampled indiscriminately with regard to *colour*. But perhaps not with regard to location (top or bottom of urn). In the example, the twins were sampled indiscriminately with regard to presence of an X chromosome (they both possessed at least one), but perhaps not with regard to location (higher or lower on the ridge). One might contend that if there has been sampling at all, then there *must* be some properties in virtue of which some were taken out, and others left in. The questions then concern (a) the nature of such properties, and (b) the role they play in (i) evolution and (ii) evolutionary theory.

[10] A. Rosenberg, A. (1978), 'The Supervenience of Biological Concepts', *Philosophy of Science* 45:368-386.

[11] Mills and Beatty (1979) [note 6].

[12] I am here assuming that the/an *epistemic* interpretation of probability is correct, that is, that probability claims are judgments as to the likelihood of a given event in light of evidence we have of past events of this sort. Probability claims, on this view, are always relevant to a body of evidence known by one or more inquirers. On a *propensity* interpretation of probability, according to which certain events have a probability of occurring quite independently of any evidence had by any inquirers, it could be assumed that the two twins had different probabilities of being struck my lightning before the momentous train of events that transpired that day on the mountain began. We could then say that the twins differed in their (objective) propensities for being struck by lightning, and hence differed in fitness with regard to lightning strikes. The twin that survived had a greater fitness than the one that died, and hence this was a selection event. This way of looking at the twin case would allow one to preserve the link between fitness differences and selection. Note, however, that it would result in the same judgment of the twin case: on either interpretation of probability, this case turns out to be an example of selection.

[13] Fitness is a theoretical concept which has its use in evolutionary theory — to capture generalizations — but is not part of the causal explanation of a particular selection event. It is not the case that *A* outsurvived *B* because *A* had greater fitness. If *A* outsurvived *B*, it was because *A* possessed some set of properties, *P*, that *B* lacked. One can say that *A* outsurvived *B* because of *A's* superior fitness, but this is just a shorthand technique which leaves entirely open precisely *why A* fared better than *B*. For this one needs to identify relevant property differences. To see how selection can discriminate between organisms which have identical fitnesses, consider the following example. Organisms *A* and *B* have the characteristics listed below (numbers represent arbitrary values of components of fitness — i.e., specific prop-

erties — which together determine overall fitness):

	Organism A	Organism B
Disease Resistance	8	2
Visual Acuity	6	4
Protective Colouration	5	5
Fleetness	4	6
Social Status	2	8
Total Fitness:	25	25

In this case, both organisms have identical fitness. But suppose that a disease epidemic breaks out, killing Organism *B* (with low resistance) but sparing Organism *A* (with high resistance). I would argue that this is an example of selection, despite the lack of overall fitness differences between the two organisms. Fitness is a useful *predictive* tool, but is not a causal factor in selection events.

[14] Sober (1984) [note 6], pp. 88-96.

[15] L. Darden and J. Cain (1989), 'Selection Type Theories', *Philosophy of Science* **56**:106-129.

[16] This is not to deny that fitness differences may be *conducive* to (or correlated with) selection events. The greater the fitness differences among organisms, the greater the chance that they differ significantly in underlying properties. The greater the difference in underlying properties, the greater the potential for selection. The important point to note here is that it is underlying property differences, not fitness differences, that are causally significant with respect to selection.

[17] Mills and Beatty (1979) [note 6], p. 268.

[18] Some recent studies on body size as an ecological and evolutionary factor include: R.H. Peters (1983), *The Ecological Implications of Body Size* (Cambridge: Cambridge University Press); W.A. Calder III (1984), *Size, Function, and Life History* (Cambridge, Mass.: Harvard University Press);

K. Schmidt-Neilsen (1984), *Scaling: Why is Animal Size So Important?* (Cambridge: Cambridge University Press); M. LaBarbera (1989), 'Analyzing Body Size as a Factor in Ecology and Evolution', *Annual Review of Ecology and Systematics* **20**:97-117.

[19] Elsewhere [T. Shanahan (1990a), 'Evolution, Phenotypic Selection, and the Units of Selection', *Philosophy of Science* **57**:210-225] I have argued at length that selection can operate on properties that may be quite temporary in their duration. Position in a dominance hierarchy, holding of a territory, and possession of mates are all highly significant factors determining reproductive success. Yet each of these can change in value many times during the life of an individual. Such characteristics may not reflect any underlying disposition for dominance, resource holding power, or sexual charisma, but rather may simply be a function of *history*: Whoever stakes out a territory (or builds a harem) first may enjoy a competitive advantage simply in virtue of being *first*. Displacement from the position of advantage and later reintroduction often results in a disfavored position. The point is that characteristics need not be stable properties of an individual to be highly significant for selection.

[20] A beautiful example of nonheritable phenotypic differences that have great selective significance: Sterility or fertility in Hymenoptera is determined by environmental conditions during growth (e.g., kind of food given), and is mediated physiologically by hormonal titers. See C.D. Michener (1974), *The Social Behavior of the Bees* (Cambridge: Harvard University Press). As Wcislo (1989, p. 157) points out, "The feedback relationships between behavior and demographic factors, and social organization and life-history traits, imply that social structure determines which reproductive opportunities will be available to individuals" (W.T. Wcislo [1989], 'Behavioral Environments and Evolutionary Change', *Annual Review of Ecology and Systematics* **20**:137-169). For more on this point, see S.A. Altmann and J. Altmann (1979), 'Demographic Constraints on Behavior and Social Organization', in I.S. Bernstein and E.D. Smith (eds.), *Primate Ecology and Human Origins* (New York: Garland), pp. 47-63, and T. Shanahan (1990b), 'Group Selection and the Evolution of Myxomatosis', *Evolutionary Theory* **9**:239-254.

[21] According to Sober, "The sort of environment an organism inhabits is part of its phenotype" (Sober 1984 [note 6], p. 119). But if so, then how can an organism interact with its environment? At most an organism can interact with part of its own phenotype! If the environment is a part of the phenotype, then selection is impossible, since selection is an interaction between phenotypic properties and critical factors in the environment.

[22] The determination of the relevant environment for a given selection event is conceptually as well as empirically problematic. Environments cannot be distinguished along sharp boundaries. In any case, there is (as yet) no completely non-arbitrary way to individuate environments, so claims that the organisms occupy the same or different environments must necessarily be inconclusive. Fortunately, selection does not require a common environment. What is required for selection is rather common "critical factors" in the environment, i.e., a common selective *agent* (Darden and Cain, 1989 [note 15]). Selection is not an interaction between an organism's fitness and its environment. Just as not all components of an organism's fitness are relevant to a selection event, so too not all components of an organism's environment are relevant. The "environment," like "fitness," is causally inert.

[23] A word of clarification: Strictly speaking, it is not that there is no biological distinction between selection and drift, because if one of these processes is more likely to lead to adaptations than the other, then there *is* a biological distinction. Rather, there is no *hard and fast* biological distinction: the two processes lie on a continuum. What differentiates selection and drift is not the nature of the events transpiring *per se*, but the kinds of *effects* or *results* one can expect from the process when different kinds of conditions obtain, e.g., whether selection takes place on widely exemplified properties, whether it is in a consistent direction, whether it results in adaptations, etc. In a previous discussion of selection and drift [T. Shanahan (1989), 'Beatty on Chance and Natural Selection', *Philosophy of Science* **56**:484-489.] I argued that selection and drift are clearly distinct, and are distin-

guished on the basis of whether sampling is on the basis of fitness differences or not. I now think that this is mistaken, or at least too simple. Evolutionary theory distinguishes selection and drift in terms of whether or not fitness differences are thought to be causative in differential biological success. But strictly speaking fitness differences have no causative power, and thus this way of distinguishing selection and drift as distinct processes *in nature* fails.

[24] Sober (1984) [note 6, p. 115] argues that "Separating selection and drift yields concepts that are needed to mark important biological distinctions." As I understand his argument, two populations may be characterised by identical sets of selection coefficients. Yet, if they differ in size they may experience quite different evolutionary careers. (In a smaller group the chances of random fixation of genes is greater.) On this view, the concept of "drift" is a way of capturing the importance of population size for evolutionary change. I would interpret this claim as follows: "Drift" is a concept used to fill in the space left between the predictions of abstract theory and the facts of concrete biological reality. Selection coefficients are educated guesses about the likely effects of certain properties in specified environments. When factored into a population genetics equation which assumes infinite population size, a prediction can be made about the change in gene frequencies in a population. But as selection coefficients are at best guesses (based on past correlations between properties and effects), they can be off the mark in actual biological scenarios. Thus, the concept of "drift" is introduced to account for changes in populations that deviate from those expected on the basis of selection coefficients. My claim is that the events described as "drift" are not different in kind from events described as "selection," but differ only in that drift events are not predictable to the extent that selection events are. Drift events are, by definition, the residue left over when populational changes fail to accurately reflect selection coefficients. Selection and drift are distinguished in theory even though they are not entirely distinct in nature.

[25] Such properties will consequently not be of great interest to many evolutionary biologists. A distinction needs to be made between those entities which function is the process of selection, and those which

function in processes of adaptation. Those biological entities that are the objects of phenotypic selection are units of selection. The objects of natural selection (which requires heritability) are units of adaptation. Units of adaptation are always units of selection, but units of selection are not always units of adaptation (Shanahan 1990a [note 19]).

[26] Rosenberg (1988) suggests two alternative interpretations of drift: (1) Drift is a cover for unknown nonevolutionary (i.e., non-adaptational, non-selective) forces. (2) Drift is a cover for selective (i.e., adaptation-producing) forces of which we are ignorant. Rosenberg seems to prefer (1). My position is neither (1) nor (2), but a third (hybrid) position: Drift is a cover for selective forces of which we are ignorant [as in (2)], but it is also non-adaptational [as in (1)]. "Drift" is a term used to cover selective events which have a low probability of leading to adaptational change. "Drift" refers to non-adaptational selective events.

[27] According to Pierre Laplace, "We regard a thing as the effect of chance when it offers to our eyes nothing regular or indicative of design and when we are moreover ignorant of the causes which have produced it. Thus chance has no reality in itself; it is only a term fit to designate our ignorance concerning the manner in which the different parts of a phenomenon are arranged among themselves and in relation to the rest of Nature" (P.S. Laplace, 'Memoire sur la probabilite' des causes pour les evenements', *Oeuvres completes*, **VIII**, 27-65; quoted in K.M. Baker (1975), *Condorcet: From Natural Philosophy to Social Mathematics* (Chicago: University of Chicago Press).

[28] When one is considering something like the origin of *Homo sapiens*, which occurred only once in the universe, one needs both general principles *and* plenty of specific details pertaining to early proto-hominid environments, etc. In a case like this, one might well choose to know everything there is about the origin and evolution of this one species, rather than settle for general principles that apply to *Homo* and lots of other primate groups as well.

[29] And scientific values determine scientific practice. As Rosenberg (1988, p. 189) points out, "The question of whether evolutionary

phenomena are statistical or not, is a different one from the question whether *our best theory* of these phenomena is unavoidably statistical" (p. 188). If it turns out that the phenomena are deterministic but that we frame our theory in statistical terms because doing so is pragmatically expedient, "then the best theory we can frame about evolution will turn out to be a *useful instrument*, but not a complete account of evolution itself" A. Rosenberg, A. [1988], 'Is the Theory of Natural Selection a Statistical Theory ?' in M. Matthen and B. Linsky (eds.), *Philosophy and Biology* (Calgary: University of Calgary Press), pp. 187-207.

[30] Sober (1984) [note 6], p. 117.

[31] Sober (1984) [note 6], p. 134.

[32] According to the Hardy-Weinberg Equation, in an infinite population drift (sampling error) is ruled out. At the other extreme, in a "population" of two organisms, one might say that whatever happens to these is, according to the theory, a case of drift (sampling error). Predictive power of the theory is proportional to population size: The larger the population, the less potential for sampling error; the smaller the population, the greater the potential for sampling error. Explanatory accuracy is inversely proportional to population size: The smaller the population, the greater the potential for determining precisely why a given sampling event occurred. Predictive power and explanatory accuracy are inversely related in evolutionary biology. Rosenberg (1988) [note 29] makes a similar point, when he points out that usefulness and realism are inversely related.

[33] S.J. Gould, S.J. (1989), Wonderful Life: The Burgess Shale and the Nature of History (New York: W.W. Norton).

Timothy Shanahan,
Department of Philosophy,
Loyola Marymount University.

SECTION THREE

THE DEVELOPMENTAL SYSTEMS APPROACH

Russell Gray

DEATH OF THE GENE: DEVELOPMENTAL SYSTEMS STRIKE BACK

"Story-telling is a serious concept, but one happily without the power to claim unique or closed readings."[1]

Evolutionary biologists tell stories. Our epic narratives sweep from diatoms to dinosaurs, from hens' teeth to horses' toes, and from protein polymorphisms to primate societies. Two philosophers of biology[2] have recently noted that there are usually two types of story: a traditional and an alternative form. The traditional type of story is told in terms of the differential survival and reproduction of individuals, while in the alternative the fundamental unit is the gene. In their paper, "The Return of the Gene", Kim Sterelny and Phillip Kitcher defend genic selectionism against several objections raised by its critics. Central to these arguments are conflicting views of the roles of genes and environment in development and evolution.

In this paper I wish to explore the implications of a recent radical reformulation of the nature/nurture debate for the types of evolutionary narratives we tell. I will begin by detailing some objections that comparative psychologists and biologists have previously raised to the division of the phenotype into innate and acquired components. I will then consider some more recent possible resolutions to the nature/nurture debate and outline the major tenets of a constructionist, developmental systems perspective. In the following sections I will consider ways in which commitments to outmoded dichotomous views of development constrain both sides of the units of selection debate. I will focus on the ways in which the views of G.C. Williams,

P. Griffiths (ed.), Trees of Life, 165–209.

Richard Dawkins, Elliot Sober, and Sterelny and Kitcher all entail a commitment to a dichotomous view of development and suggest ways in which their ideas may be reworked from a developmental systems perspective. I will conclude by exploring the new narrative fields that are opened up once these commitments are exposed and abandoned.

1.INNATENESS AND THE PROBLEM OF GENETIC DETERMINISM

In the study of animal behaviour there is a long history of intense and productive debates over the concept of innate behaviour and the related problem of genetic determinism[3]. Today both the concept of innate behaviour and genetic determinism are rejected by most developmental biologists and psychologists[4]. To understand why, and to lead us on to more recent views, I will briefly review four common objections to the division of behaviour into innate and acquired components[5].

1.1 MULTIPLE MEANINGS.

The first, and apparently most trivial objection to the concept of innate behaviour, is that the term is often used in a multiplicity of non-equivalent ways. Bateson[6] noted that there are at least seven different senses in which the terms "innate" or "instinctive" might be used: "present at birth; a behavioural difference caused by a genetic difference; adapted over the course of evolution; unchanging throughout development; shared by all members of a species; ... not learned," and " ... a distinctly organised system of behaviour driven from within." I should emphasise that the problem of multiple meanings is only apparently trivial. If the problem were merely one of sorting out the confusion caused by using a word with many different meanings then the issue would have been cleared up years ago with a few new terms and redefinitions. More substantive problems arise because people often act as if evidence for one definition of innateness implies another. This is clearly not the case. Bateson illustrated this well with the following example. In his classic paper on "The ontogeny of an instinct" Hailman[7] found that although young laughing gull chicks are able to peck at their parents bills for food without prior practice or the chance for observational learning (thus satisfying at least three definitions of innate behaviour - adaptive, present at birth, not learnt), this behaviour was modified by subsequent learning (contradicting another definition of innate behaviour). The underlying reason why people often slide between different senses of innateness is that these senses are all linked to a dichotomous view of development. Innate behaviour is seen as being determined by internal, inherited factors while acquired behaviour is driven by the environment. I will address some reasons why this dichotomous view of development is misleading at the end of this section.

1.2 IMPLICIT PREFORMATIONISM.

Implicit in the concept of innate behaviour is the idea that behaviour is somehow present in either a latent or coded form in the genes. Such a view is implicitly preformationist as it implies that the behaviour exists prior to developmental processes and interactions rather than actually being constructed epigenetically through ontogeny. In its most extreme form this point of view is obviously wrong. Genes do not contain behaviour. Diluted forms of this view (e.g., genes contain a blueprint for behaviour or a programme for behaviour) appear more reasonable as they allow a role for developmental processes. However, they are still, at heart, preformationist. They are preformationist rather than epigenetic because they assign only a secondary role to these developmental processes of "expressing" or "translating" the underlying blueprint or programme. The basic form or structure still preexists these developmental processes. (This is reminiscent of Aristotle's form/matter distinction. The genes are the source of form or design and the environment provides matter or material support for the implementation of this design). As numerous developmental researchers[8] have stressed, this view results in a pseudo-explanatory account of the way in which innate behaviour actually develops. When asked to explain how innate behaviour develops proponents of innate behaviour merely reassert that the trait is innate. The actual development of so-called innate behaviours is not studied, analysed nor explained. Their existence is simply taken as given.

1.3 DEPRIVATION OR SELECTIVE REARING?

One of the main criteria ethologists such as Lorenz used to divide behaviour into innate and acquired components was the outcome of what were termed "deprivation experiments". The logic of the deprivation experiment seemed simple enough (given a dichotomous view of development). The aim was to remove the relevant environmental sources of information so that only those of the genes remained. Animals were therefore raised from an early age in an environment that deprived them of the possibility of learning, practicing or observing the behaviours under question. If the behaviour still developed then it must be innate, whereas if it failed to develop it must be learnt. Lorenz insisted on this dichotomy, attributing intermediate cases to bad rearing conditions disrupting genetically determined maturation, poor experimental design or the misclassification of behaviour.

Numerous people have pointed out that the interpretation of deprivation experiments is rather more problematic than this.[9] The deprivation experiment suffers from the same problems as all other attempts to prove a universal negative statement. The definition of innate behaviour as behaviour that is not learnt requires the experimenter to eliminate all possible potential sources of environmental information. Obviously it is impossible to raise an organism in the absence of an environment or without experience. The logic of the deprivation experiment therefore depends critically on being able to distinguish between environmental conditions that merely provide support for developmental processes and those that provide specific relevant information. The deprivation experiment allows the former but aims to eliminate the latter. Unfortunately, in practice this distinction has proved unworkable.[10]

Most current researchers would now reinterpret deprivation experiments as selective rearing experiments. The logic of this interpretation is quite different from that of the deprivation experiment.

Instead of being designed to reveal whether the behaviour is innate or learnt, the selective rearing experiment tests whether a specific environmental factor is developmentally relevant. Only a positive outcome can be interpreted (i.e., changing the environment produces a difference in behaviour). In this case a factor that influences development has been identified. If there is no change in behaviour then little can be said other than that the environmental factor did not change the animal's behaviour (at least at that specific time of its life).

1.4 DEVELOPMENTAL DICHOTOMIES.

Underlying all the problems outlined so far is a dichotomous model of development. In this model, illustrated in Figure 1, there is a simple correspondence between developmental inputs and behavioural out-

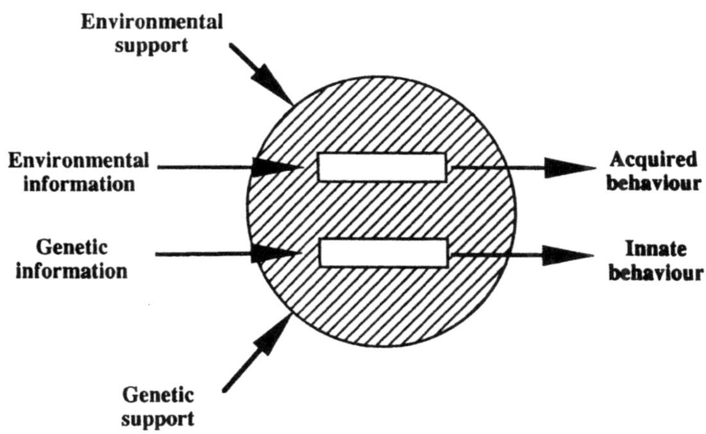

Figure 1 : The dichotomous view of development in which there is a simple relationship between developmental inputs (genes and environment) and two kinds of behaviour (innate and acquired). This view also distinguishes between developmental information and background developmental support. *

comes. Environmental information produces acquired behaviour and genetic information produces innate behaviour. It is this dichotomous view of development that causes people to illegitimately conclude that because a behaviour is present at birth it required no experiential factors to develop, and/or that it cannot be modified by subsequent experience. This dichotomous view also causes people to reason that if one source of developmental information is cut off (no learning), then what still develops must be due to the other (the genes). The complexity of developmental processes undercuts such a simple classification. Even at the molecular level there is no simple correspondence between the sequence of DNA base pairs and the functional activity of the proteins they are claimed to "code" for. Although the nucleotide sequence does specify the primary structure of a protein (its sequence of amino acids), it is the tertiary structure of the protein that determines its function, and this depends on a range of non-genetic chemical and physiological factors inside the cell.[11]

At the behavioural level the picture is no less complex. Upon experimental investigation it has been found that, far from developing autonomously, the forms of behaviour that might have been termed "innate" or "genetically determined" (behaviour that is species-specific, adaptive, present at birth, difficult to change etc.), all require experiential inputs in their development. These inputs are not just secondary and supportive but *"positive, informative* and *constructive."*[12] This, of course, is not to say that all behaviour is learnt but rather that the categories "innate" and "learnt" are misleading. Experiential inputs into development are far broader than just direct learning. Often the effects of experiential inputs are often more subtle, indirect and less specific than straight forward learning.

Gilbert Gottlieb's classic series of experiments on the development of ducklings' preferences for the maternal call of their own species illustrates this very well.[13] Soon after hatching young ducklings show a clear preference for the maternal call of their own species. This response would satisfy many of the criteria often used to classify a behaviour as innate. It is species-specific, adaptive, present at birth and appears to develop without the possibility of prior learning. What Gottlieb discovered was that if the ducklings were devocalised *in the egg* then they would no longer show the same clear preference for the call of their own species. Devocalising the ducklings in the egg deprived them of the opportunity to make calls and thus to stimulate the development of their auditory system. This *prenatal* self-stimulation was therefore necessary for the development of the *postnatal* preferences. However, in hearing its own embryonic call the duckling is not learning the character of the maternal call (the calls are quite different). Instead this non-obvious, internally generated experience was stimulating the development of the auditory system that would later be used to detect the species-specific maternal call.[14]

If, on close inspection, the category "innate behaviour" must expand to include all sorts of behaviour influenced by experience, so then must "learnt behaviour" be dependent upon internal factors. Learning cannot take place without a sensory system to pick up the environmental stimulation. Learning, and the wider effects of experience, are contingent upon genotype. Dudai and co-workers, for example, found that the ability of fruit flies to learn olfactory discriminations varied with genotype[15]. Similarly, Sackett and co-workers found that although early social isolation had a profound negative effect on the development of individual, social and exploratory behaviour of rhesus macaques, its impact on pigtail macaque and crab-eating macaque development was quite different. In spite of being raised in social isolation, pigtail macaques developed reasonably normal individual and exploratory behaviour and crab-eating macaques developed reasonably normal exploratory and social behaviour.[16] The effects of social isolation are thus contingent upon the species being studied. So upon developmental analysis the two categories "innate" and "learnt" behaviour expand, interramify and coalesce, and thus cease to be meaningful. *All* phenotypes are the *joint* product of internal and external factors. Development is one process requiring many inputs linked together by complex, non-linear, dynamic systems. It cannot be realistically chopped into two.

2. NATURE AND NURTURE TODAY

For researchers in the field of behavioural development the largely historical analysis of the nature/nurture debate I have presented so far will, I guess (or hope), seem familiar and fairly uncontroversial. Nowadays it seems that everybody is an "interactionist". Unfortunately, frequent incantations of the word "interaction" and the familiar homily that, "of course all phenotypes depend on both genes and experience", have not been enough to drive away the ghost of dichotomous views of development. In fact, as has recently been pointed out,[17] the dichotomous view is alive and well, albeit living under assumed names. So instead of referring to innate or genetically determined characters, researchers will use weaker terms such as genetic blueprints, programmes, tendencies or predispositions. Instead of partitioning behaviour into inherited and acquired components, they attempt to separate the effects of maturation from experience, or phylogenetic from ontogenetic sources of information. Although these terms may seem more reasonable, as Oyama and Johnston have emphasised, they suffer from all the difficulties outlined above. Diluting the poison doesn't make the problem go away.

One way of conceptualising the contributions of genes and environment to development that appears to avoid these problems is provided by the old, but increasingly fashionable, "norms of reaction" concept.[18]

"The norm of reaction for a genotype is a list or graph of the correspondence between the different possible environments and phenotypes that would result ... Rather than characterising a genotype by a single phenotype or a single "tendency", the norm of reaction describes the actual relation between the environment and the phenotype for the genotype in question."[19]

Figure two is an example of some norms of reaction - cuttings from seven different individual plants of the Californian herb the milfoil (*Achillea millefolium*) grown at three different altitudes.

These norms of reaction (and reaction norms in general) make the following facts about development very clear. First, specifying the genotype or the environment alone is insufficient to specify the phenotype. Both must be specified in order to specify the height and branching form of the herb. Phenotypes are indeed a joint product of genes and experience. Second, it is not possible to assign causal

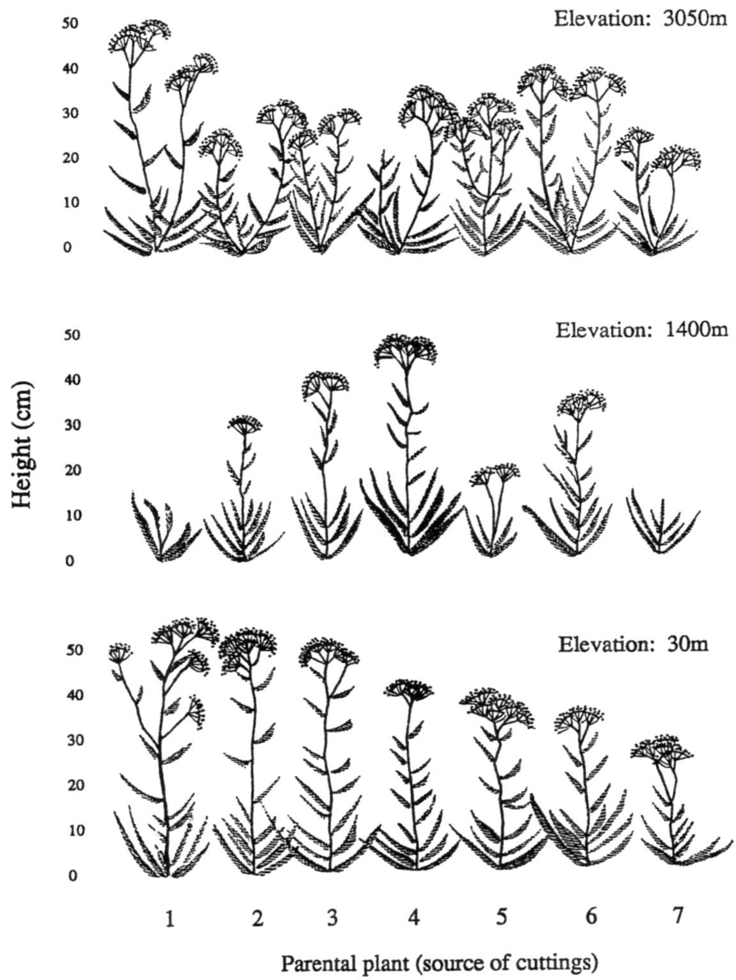

Figure 2: Norms of reaction for the herb *Achillea millefolium*. **

primacy to either factor. The effect of the genotype is contingent upon the environment, and the effect of the environment is contingent upon genotype. Third, the relationship between genotype and environment revealed by norms of reaction is not the type that would be predicted if two developmental factors were simply added together. The relationship is strikingly non-additive. The cuttings in Figure 2 are arranged so that at the lowest altitude the tallest plant is on the left and the shortest on the right. At the higher altitudes this order is completely changed. The cutting from plant number one is the tallest plant at the lowest altitude, almost the shortest at the middle altitude, and the tallest, but branching at the base, at the highest altitude. Lewontin concludes his discussion of these norms of reaction with the comment that, "No notions of determination, or tendency, or capacity have any meaning for describing the relations between genotype and phenotype. The phenotype is the unique consequence of a particular genotype developing in a particular environment".[20]

However, all is not yet sweetness and light. Although the norms of reaction concept is an advance on the views discussed earlier, from a developmental perspective it has a number of limitations as Ho[21] has pointed out. First, it presents a very static view of development. A more adequate representation of development would represent norms of reaction as a range of life-history trajectories over a range of environments. Second, the environment itself changes over development. The environment of a tadpole is different from that of a frog. Similarly, the environment of a human infant is different from that of an adult. A better picture still would include a range of life-history trajectories over a range of codeveloping environments. As an accurate representation of development even this view is still limited. The organism not only reacts to the environment, it acts on it. Earthworms modify the soil, birds build nests, beavers construct dams and we warm the planet.[22] Any truly adequate representation of development must capture this dynamic.

3. CONSTRUCTIONISM - A RADICAL REFORMULATION OF THE NATURE/NURTURE DEBATE

The exasperated reader must now be wondering if any conceptualisation can possibly avoid all the pitfalls that lurk in developmental debates. Have hope. I believe there is a view that is not implicitly dichotomous, preformationist and genetically determinist. This view involves a radical reformulation of our view of developmental processes.[23] I will term this new view constructionism. The constructionist view of development is not attributable to any one person or group. It is a label for a perspective that has developed, often quite independently and to different degrees, at various sites around the globe.[24] The central tenets of this perspective could, perhaps, be formalised as follows:

3.1 JOINT DETERMINATION.

All phenotypes, be they physiological, morphological or behavioural, are jointly determined by both genes and the developmental context. This truism means that it is not possible to assign causal primacy nor to dichotomise developmental causation into internal and external components.[25]

3.2 RECIPROCAL CONTINGENCY.

Development is a contingent, conditional process. This claim builds on the basic truism of joint determination. It notes that the effects of both genetic and environmental differences are contingent on the context in which they occur. Thus the effects of changes in the environment will depend upon the organism's genotype and vice versa. Not only are they reciprocally contingent but, as Stent, Lewontin, and Nijhout stress,[26] they are temporally contingent. The impact of an environmental factor will vary depending on the developmental state of the organism and, reciprocally, the effect of a gene being activated will depend on the state of the rest of the developmental system. Lewontin[27] gives the following example. If a fruit fly egg is exposed to ether fumes about three hours after it has been fertilised, the adult fly will develop a rudimentary additional pair of wings (the "bithorax" phenocopy). If it is exposed to ether after this time the bithorax phenotype will not occur.[28] The reciprocal and temporal contingency of developmental events frustrates any effort to represent developmental causation as the simple addition of a genetic and an environmental vector. Developmental events do not have constant additive effects. Developmental causation must therefore be conceptualised in system rather than vector terms.

3.3 CO-DEFINITION AND CO-CONSTRUCTION.

In development, internal and external factors are co-defining and co-constructing.[29] While it may still be convenient to talk of "internal" and "external" factors, in reality "internal" and "external" factors are not independent variables and do not exist, in any meaningful way, in isolation from one another. Thus a gene can only be functionally defined in a specific developmental context. "A gene for blue eyes" is only "a gene for blue eyes" when it is embedded in an appropriate developmental context. Similarly, an organism's environment is best defined in organism-referent terms. Lewontin gives the following example. For a bacterium in a pond Brownian motion is a major environmental feature and gravity is not, while for a heron in the same pond quite the opposite is the case. Or, to use the example of the impact of an ether shock on fruit fly development outlined above, an ether shock is only an ether shock when there is a developmental system to shock that is sensitive to it. Any meaningful description of environmental resources or information must also be implicitly organism-referent. What are resources to one organism may be irrelevant to another. Any statement about resources limiting a particular population must be contingent upon a given range of actions by that population. The information contained in the ambient flow of light or in the vibration of air molecules will be contingent upon the organism perceiving that information. Not only are different organisms sensitive to different frequencies, they pick up different information from the same physical stimulus. Light reflected from a tree may afford a potential food source to one organism, shelter to another and a good position to sing from for another.

Not only are "internal" and "external" factors co-defining they are also co-constructing. An organism's environment plays an essential role in the construction of its phenotype and, reciprocally, the organism selects and modifies (constructs) this environment out of the resources that are available. This "external" construction may range from the dramatic building of nests, hives, dams, computers and international banking systems to more subtle forms of construction such as the shedding of leaves and bark or the release of allelochemicals.

3.4 CONSTRUCTION, NOT TRANSMISSION.

All phenotypes are constructed, not transmitted. This is, perhaps, the most important tenet of the constructionist view. Not only are traits not directly transmitted across generations, nor are blueprints, potentials, programmes or information for the traits. Instead, all phenotypes must develop through organism-environment transactions. This thoroughly epigenetic view of development is the reason for the apparently obscure title to Susan Oyama's book, "The Ontogeny of Information"[30]. In the book she argues that developmentally meaningful information does not pre-exist developmental processes. Developmental information is not *in* the genes, nor is it *in* the environment, but rather it develops in the fluid, contingent *relation* between the two. Information itself has an ontogeny.

At first encounter, this view may seem a little odd, so perhaps the following analogy might help. DNA is a bit like a literary text. Just as rows of letters linked into smaller and larger units have no intrinsic meaning, but rather their meaning is constructed in the context of a reader of the appropriate culture and experience, so DNA sequences have no fixed, essential meaning. Information is constructed in the relationship between text and context, and thus as deconstructionist literary theorist playfully emphasize, is constantly open to re-reading. If this analogy is a little too French for some readers' taste, then perhaps a computational analogy might suffice. DNA is like a string of binary machine code. The "code" only has meaning when run on a machine with the appropriate hardware and operating system. This analogy is not quite so good, however, as it potentially opens up the way for a form/matter type distinction and thus leads us back to a dichotomous view of development. Numerous other analogies are possible. Orchestral productions and plays are obvious candidates. The problem with all these kinds of human analogies is that it is very tempting to conceive of them in dichotomous terms where a distinction is drawn between the original or intended design and the secondary production of that design. (For an analogy that does escape from this temptation see tenet seven.)

3.5 DISSIPATED CONTROL.

As information is not localised in the genes, nor in any particular entity, the control of development does not reside in the genome. Instead it

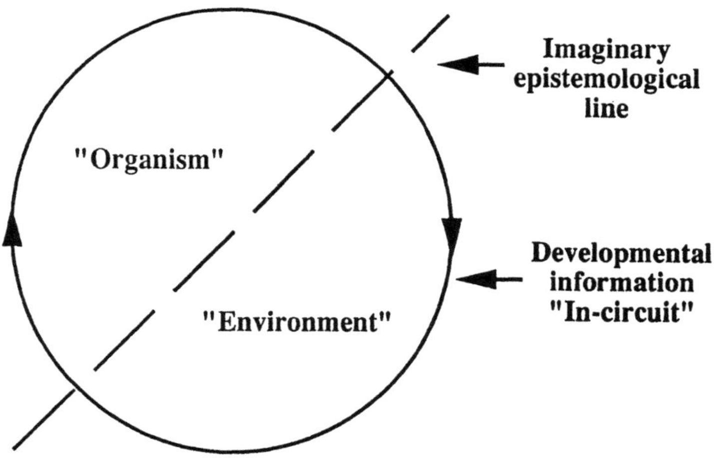

Figure 3: A slice through the organism-environment system at one point in time.[31]

is diffused, dissipated or disseminated across the entire developmental system. Genes are not "hegemonic" controllers of development (the master molecule myth) they are just one interactant in a field of context-dependent difference makers. I have tried to capture this view in Figure 3 with a slice through the organism-environment system at one point in time. In such a system developmental information is constantly "in circuit" rather than localised in either internal or external factors.

3.6 EXPANDED INHERITANCE.

Developmental resources or interactants are inherited, and the set of these factors is much wider than just genes.[32] This tenet follows as a consequence of the claim that all phenotypes are constructed not transmitted. If traits are not transmitted then the obvious question that arises is, "just what exactly do organisms inherit?" The constructionist answer to this question is that organisms inherit a large set of developmental interactants (see Table 1).[33] These interactants obviously include genes and cytoplasmic factors. The set of interactants also includes chemical cues passed on from the maternal diet (via fetal olfactory conditioning and maternal milk in rodents, and chemical traces left on the egg in insects). Gut micro-organisms can also be passed on across generations. Social traditions can play an important role in the reconstruction of feeding methods, migration routes and schooling sites, home ranges and territories, reproductive sites, dominance rank and birdsong across generations. As Vygotskian developmental psychologists emphasise human infants inherit a structured social environment.[34] At the most general level organisms may also inherit a large range of physical conditions (e.g., altitude and climate) and biological factors. If "earth and life evolve together" (share a common history of differentiation) as panbiogeographer Leon Croizat claimed, then the inheritance of these factors may be much more pronounced than had previously been thought.[35] Ecological associations and communities and their landscapes might function as co-evolutionary units.

To argue for an expanded view of inheritance is not to argue for

Type of inheritance	Reference[33]
Cytoplasmic factors	Cohen (1979), Ho (1984)
Chemical traces from parental foraging	Galef and Henderson (1972), Corbet (1985), Hepper (1988)
Gut micro-organisms	Mattson (1980), Jones et al (1981)
Social traditions	
Feeding methods	Fisher and Hinde (1949), Norton-Griffiths (1968)
Migration routes and schooling sites	Van Denburgh (1914), Helfman and Schultz (1984)
Home ranges and territories	Neal (1948), Carrick (1963), Jolly (1972), Woolfenden and Fitzpatrick (1978)
Reproductive sites	Harris and Murie (1984)
Dominance rank	Cheney (1977), Horrocks and Hunte (1983)
Song	Jenkins (1978)
Other features of the environment e.g. geographic range	Croizat (1964), Rosen (1978), Jablonski (1987)

Table 1: "Extra-genetic" inheritance

a dual inheritance system (e.g., biological and cultural).[36] That would lead us back to a view that there are two sources of developmental information, and thus back to a dichotomous view of development. A thoroughly constructionist view of development would emphasise that, just as the phenotypic effect of a genetic interactant depends on the developmental context, so the effects of extra-genetic inheritance will be conditional on the activities of the organism.

3.7 INTERACTIVE CONSTANCY.

Developmental and evolutionary constancy are due to the constancy of patterns of interaction rather than the constancy of any single factor. Any adequate view of development must explain the remarkable developmental constancy that exists across generations (humans rarely have two heads). Dichotomous views of development achieve this by appealing to either the constancy of a preformed genetic programme or to the constancy of the environment. However, just as in baking a cake using the same ingredients is not enough to guarantee a constant outcome (they must be put together in the right way), so too in development it is the *constancy of process* that counts.[37]

Stent[38] captures this idea with what I think is a wonderful and (at last) appropriate analogy. Development is like an idealised process of ecological succession. Ecological succession proceeds in a relatively orderly manner due to the regular sequence of interactions between organisms and their environment. Consider the following (highly idealised) succession scenario. Bare rock is exposed as a consequence of volcanic or glacial action. After the rock has weathered sufficiently to allow the formation of some soil, lichens colonise the surface and accelerate the formation of soil. These interactions create a suitable environment for grasses and herbs to colonise and they eventually replace the lichens. Larger shrubs and trees follow the colonisation of this environment by grasses and herbs providing a suitable habitat for herbivorous animals. Once herbivores are present then omnivorous and carnivorous organisms can then colonise the developing community. Ecological succession does not require an ecological blueprint or programme to control its development. Each stage creates the conditions necessary for the next. The process is *self-organising*. The absence of an underlying programme controlling development does not mean that the transgenerational *reconstruction* of the phenotype is free to proceed in any direction. At any point in development it is constrained by the current state of the organism (which is the product of past organism-environment transactions) and its current environmental context. Phenotypic reconstruction is reciprocally constrained.

3.8 NATURE AND NURTURE REFORMULATED.

Nature is the *product* and nurture is the *process*. Susan Oyama[39] reformulates the relation between nature and nurture by noting that the usual error made by dichotomous views of development is to see nature (e.g., "human nature") as genotypic rather than phenotypic - a source of developmental information rather than an outcome of developmental processes. She argues that as an organism's nature (its characteristics at any point in time) is a phenotypic construction, this nature must be dynamic and multiple rather than fixed and unitary. From this perspective nurture is not a second source of developmental information, but rather the developmental processes that construct the phenotype from both "internal" and "external" resources.

3.9 WHAT IS EVOLUTION?

Evolution is change across generations in the distribution and composition of populations of developmental systems. The traditional view of evolution defines it as "change in gene frequencies".[40] Numerous authors have recently argued that a constructionist view of development and an expanded view of inheritance require a developmental systems rather than a gene-centred view of evolution.[41] From this perspective just as it is not organisms that develop, but organism-environment relations, so it is not populations that evolve, but rather population-environment relations.[42] A change in gene frequency or a change in any of the extra-genetic factors listed in Table 1 can cause a transgenerational change in the distribution of phenotypes in a population.[43] To regard only genetic changes as evolution is to give genes prior ontological status over other context-dependent difference makers, and so return to the distinction that is the root cause of the nature/

nurture dispute. Furthermore, it is to ignore what evolutionary biologists are really attempting to explain. As Lewontin remarked,

"Population geneticists, in their enthusiasm to deal with the changes in genotype frequencies that underlie evolutionary changes, have often forgotten that what are ultimately to be explained are the myriad and subtle changes in size, shape, behaviour and interactions with other species that constitute the real stuff of evolution."[44]

No doubt it might be claimed that as these extra-genetic changes are easily reversible they must ultimately become fixed in the genome to constitute a major factor in evolution. There is, however, no a priori reason to assume that extra-genetic changes are more or less reversible than genetic changes.[45] Back-mutations, the random elimination of alleles, a change in the developmental context, and a change in selection "pressures" could all reverse the effects of genetic changes. Extra-genetic changes such as a shift from a marine to a terrestrial environment, a change in host plant type, or the geological fragmentation of an ecological community may persist for millions of years. The effects of sensitive periods to environmental stimulation, frequently found in early ontogeny, could assist the establishment of relatively irreversible changes in habit or habitat without the necessity of genetic changes. For example, each generation could be imprinted on the habitat it was born in. From this perspective evolution can take place *without* a change in gene frequencies. In reality, of course, it is likely that genetic and extra-genetic changes will occur together. The point remains, however, that extra-genetic changes can persist for relatively long periods of evolutionary time and thus are a significant part of the "real stuff" of evolution. This is important as it frees proponents of developmentally based accounts of evolution from having to demonstrate that developmental changes are "genetically assimilated" or that Weismann's barrier is violated. This is a trap because it endorses the orthodox premise that evolution ultimately is *just* a change in genotype frequencies. The view of heredity as an ecological process frees evolutionary biologists from this trap and enables us to begin synthesising developmental and evolutionary explanations.

4. DEVELOPMENT AND THE UNITS OF SELECTION DEBATE

4.1 GENIC SELECTION

Armed with both a new view of development and a development-centred view of evolution we are now ready to tackle the units of selection debate. Let's start with Richard Dawkins and "The Selfish Gene".[46] In his wonderful metaphor packed book, "The Selfish Gene", Dawkins built on the earlier work of Hamilton and Williams to offer a radical revision of the orthodox Darwinian view of nature.[47] Dawkins claimed that genes not individuals were the fundamental unit of selection. Central to Dawkins' gene-centred view of evolution was a distinction between two types of entities - replicators and vehicles. Replicators were entities that interacted with the environment in such a way that copies of themselves were made. According to Dawkins the gene was an obvious replicator although other units such as units of culture (memes) may also have functioned as replicators. In evolution these replicators competed through the construction of ephermeral phenotypic vehicles. The better the vehicle the more replicator copies were produced. The replicators were thus potentially immortal whereas the vehicles had to be rebuilt each generation. Both the colour and the substance of the vision is captured in the following, often cited passage.

"Replicators began not merely to exist, but to construct for themselves containers, vehicles for their continued existence. The replicators which survived were the ones which built survival machines for themselves to live in. The first survival machines probably consisted of nothing more than a protective coat. But making a living got steadily harder as new rivals arose with better and more effective survival machines. Survival machines got bigger and more elaborate, and the process was cumulative and progressive....Was there to be any end to the gradual improvement in the techniques and artifices used by the replicators to ensure their own continuance in the world? They did not die out, for they are past masters of the survival arts. But do not look for them floating loose in the sea; they gave up that cavalier freedom long ago. Now they swarm in huge colonies, safe inside gigantic lumbering robots, sealed off from the outside world, communicating with it by tortuous indirect routes, manipulating it by remote control. They are in you and in me; they created us, body and mind; and their preservation is the ultimate rationale for our existence. They have come a long way, those replicators. Now they go by the name of genes, and we are their survival machines."[48]

Dawkins' critique of the individual as the unit of selection did not stop with the reduction of individual phenotypes to replicator power. In the "Extended Phenotype"[49] he expanded the power of the genes out beyond the skin so that the behaviour of other organisms and even the physical environment could all be seen as expressions of replicator power. Thus, "...the whole world is potentially part of the phenotypic expression of a gene."[50] Here is Dawkins's typically pithy summary of his views:

"To conclude: the replicator is the unit of selection. Adaptations are for the benefit of replicators. Individuals are manifestations of the power wielded by replicators over the world in which they live. The individual body is a convenient practical unit of combined replicator power. But we must not be misled by the parochial detail. In the light of the doctrine of the extended phenotype, the conceptual barrier of the individual body wall dissolves. We see the world as a melting pot of replicators, selected for this power to manipulate the world to their own long-term advantage. Individuals and societies are by-products."[51]

Critics of Dawkins' gene-centred view were quick to take up the challenge of such a radical view. Gould, for example, accused Dawkins of the three deadly sins - atomism, reductionism, and determinism.[52] Bateson[53] patiently argued against giving special status to genes as the prime controllers of development. Certainly all this talk of genes looks and smells like genetic determinism. Not so, claimed Dawkins. Although he may have got a bit carried away with some of his metaphors in "The Selfish Gene", he argued that nothing in this gene-centred view of evolution required a commitment to a gene-centred, genetic determinist view of development.

"Bateson is worried that I seem to give 'special status' to genetic determinants of behaviour. He fears that an emphasis on the gene as the entity for whose benefit organisms labour, rather than the other way around, leads to an undue emphasis on the importance of genetic as opposed to environmental determinants of development. The answer to this is that when we are talking about *development* it is appropriate to emphasize non-genetic as well as genetic factors. But when we are talking about units of selection a different emphasis is called for, an emphasis on the properties of replicators. The special status of genetic factors rather than non-genetic factors is deserved for one reason only: genetic factors replicate themselves, blemishes and all, but non-genetic factors do not."[54]

From a developmental systems perspective there are a number of obvious problems with Dawkins' position. First, genetic factors do not replicate themselves nor do they physically persist across generations. They are replicated as part of the *reproduction* of developmental systems. Remove some part of that developmental system and genetic replication may be changed or impaired. In this sense genes are no different from any other developmental interactant.

Second, non-genetic factors *are* replicated across generations (see Table 1). It could be objected that although some of the forms of extra-genetic inheritance are indeed replicated (e.g. cytoplasmic factors and cultural traditions), others simply persist. Gravity, sunlight and climatic factors all apparently persist. However, from a developmental systems perspective the environment must be described in organism-relevant functional terms. Thus what is replicated is the role of these factors as part of the informationally salient features of certain developmental systems. As such if other aspects of the developmental system are changed then the role of gravity, sunlight or climate may be altered. Gravity, sunlight and climate are not fixed features of our own species' developmental system. If we were to take up life in space, in caves, or in Antarctica their developmental role would cease or change. It is in this sense that these factors are replicated. Each lifecycle their developmental role is replicated.

The third, and comparatively minor, terminological problem with Dawkins' view is, as David Hull has pointed out, that the term vehicle makes the phenotype seem causally passive.[55] Vehicles require agents to drive them. To see vehicles as products of replicator power is to place genes in the driving seat and thus travel back down the road to genetic determinism.

The fourth, and most fundamental problem, is that evolution cannot be glibly separated from development. Evolutionary causes are no different in kind from developmental causes. Dawkins and his fellow genetic selectionist Williams attempt to separate off evolutionary causes by separating inheritance from development. In doing so they must implicitly endorse a preformationist view of development. This can be seen very clearly in Williams' discussion of Elliot Sober's book "The Nature of Selection."[56] Williams claims that genes are the only biological entities that persist across generations and thus the only candidates to be units of selection. Obviously I would wish to draw his attention to the continuity of extra-genetic factors. In response to Sober he concedes that genetic continuity across generations is not physical identity but argues that it is the gene-as-information that persists. "A gene is neither an object nor a property but a weightless package of information that plays an instructional role in development."[57] But, as I have emphasised earlier, developmental information is not carried in preformed packages. Developmental information is not *in* the genes. Developmental information has a context-dependent ontogeny. It is constructed not transmitted. This construction process will reproduce both genetic and extra-genetic entities, however, from a developmental systems perspective the unit of replication is neither the genes nor any extra-genetic factor. The unit of replication is a relationship not an entity. Williams was on the right track with his emphasis on information. It is the reoccurrence of patterns of developmental *interaction* that leads to the *reconstruction* of the same developmental information. The unit of replication is these temporally structured patterns of developmental interactions - life history trajectories.

This view causes a little mischief with some of our traditional distinctions. It blurs Hull's distinction between replicators and interactors. Hull notes that DNA is both an interactor and a replicator.[58] From a developmental systems perspective all aspects of the developmental system are part of the replicator process and all are interactors. It is this interaction that produces replication. No single factor is a replicator. All are interactors. Interaction and replication are two roles for the same entities and processess.

4.2 FOREGROUNDING THE FOOTNOTES.

One of the most detailed analyses of the units of selection debate has been provided by the philosopher Elliot Sober.[59] Sober claims that genic selectionism is an ambiguous thesis. It actually makes two claims - a substantive claim that all adaptations exist because they benefit individual genes and a trivial claim that all evolutionary change can be represented in genic terms. He notes that when genic selectionists such as Williams and Dawkins are pressed on the substantive claim they fall back to the trivial claim. For Sober the substantive claim is about what there is *selection for* whereas the trivial claim is about what there is *selection of*. He uses the following ingenious example to illustrate this distinction.

"My young son has a toy which takes all the mystery out of this distinction. Plastic discs with circles cut out of them are stacked with spaces in between in a closed cylinder. The top-most disc contains very big holes, and the holes decrease in size as one moves down from disc to disc. At the top of the cylinder are found balls of different sizes. A good shaking will distribute the balls to their respective levels. The smallest balls end up arrayed at the bottom. The next smallest sized balls settle at the next level up, and so on. It happens that the balls of the same size also happen to have the same colour. Shaking sends the black balls to the bottom, the pink ones to the next level up, and so on. The whole cylinder (plus paternally administered shaking) is a selection machine. The device *selects for* small balls (these are the ones which pass to the bottom). It does not *select for* black balls (even though these are the ones which pass to the bottom). But when we ask after a shaking what was selected, it is equally correct to say that the black balls were selected and that the small ones were. 'Selection for' focuses on causes; 'selection of' picks out effects."[58]

Sober summarises this view by noting that we can speak about selection *for properties* and *of objects*. The trivial genic selectionism thesis that there is always selection *of genes* does not, therefore, imply that there is always selection *for genes*. According to Sober changes in gene frequency will often be shadows of higher level causal processes.

Consider the familiar example of heterozygote advantage - sickle-cell anaemia. In malaria infested regions of Africa humans that are heterozygous for the "sickling" allele (i.e., SA genotypes) are fitter than the homozygous genotypes (SS or AA). Homozygotes for the normal allele A are more vulnerable to malaria than the heterozygotes, while those that are homozygous for the sickle allele S are anaemic. Sober argues that in this situation although there is *selection of* the alleles S and A there is only *selection for* the genotype SA. In Sober's terms the SA genotype is the positive causal factor not S or A. These individual alleles are not positive causal factors because they do not have uniform causal effects. Their effects vary with the allelic context. Changes in the frequencies of the alleles S and A are thus shadows of selection acting at the level of the genotype.

So far so good. Sober's influential critique of genic selectionism has certainly helped a great deal to clarify many of the central issues in the units of selection debate. But from a developmental systems perspective what is equally revealing is the point at which this critique stops. Sober leaves unquestioned many of the traditional assumptions of evolutionary theory. The limitations of some of these traditional assumptions can be exposed by foregrounding some of the contradictions tensions and equivocations tucked away in the footnotes of the "The Nature of Selection". I would now like to focus on these footnotes and ask how Sober's analysis of the selection process might be extended by adapting a constructionist perspective.

Throughout his book Sober adopts the orthodox view that evolution is a "change in gene frequency". This position is probably just a convenient starting factor rather than a fundamental assumption. Sober appears to acknowledge this in footnote 1 on page 278 of "The Nature of Selection", "Here I continue to play along with the assumption that evolution requires a change in frequencies, although the limitations of this view have already been set forth". In an earlier footnote on page 255 he notes, "The fact that there can be extrachromosomal hereditary mechanisms should not be forgotten. Genes are not the only units of replication". So what would happen to Sober's analysis if evolution was regarded as a change in the composition and distribution of populations of developmental systems, and if extragenetic inheritance was emphasised rather than marginalised to the footnotes? I believe Sober's case against genic selectionism would be greatly strengthened. For a start even the trival claim that evolution can always be represented in genic terms would not be true. If a representational argument were to be maintained it would have to be widened to include extragenetic factors. Apart from being rather messy, and particular (the type of extragenetic factors would quite often be lineage specific), this expanded representational account would undermine the apparently unique role of genes as the only constant factor in a sea of transgenerational flux. "The Selfish Gene, Cytoplasm, Social Tradition, Species' Range ..." lacks a certain punch. Genes would quite correctly just become one of the many developmental interactions that are replicated across generations.

Sober's case against genic selectionism can be further strengthened by highlighting the apparently arbitrary limit that he places on factors that can count as positive causal factors[61] (factors that there can be selection for). Sober's discussion of the case of heterozygote advantage illustrations this well. He argues that there is *selection for* the genotype SA not for the individual alleles because the genotype has a unique causal role whereas the individual alleles do not. But the genotype does not have a unique causal role. Its causal role is context dependent in just the same way as that of the alleles. The heterozygote genotype is only advantageous in malaria infested environments. This environment may, in the sense outlined above, be inherited. So why does Sober not apply his principle of causal uniformity and include the relevant features of the environment as part of the positive causal factor? Are we once again seeing the "environment" being regarded as nice background support rather than positive, informative and constructive? As Lloyd notes, "The question of how much structure of the natural system must be specified in the model is the units of selection question ..."[62]. To suggest that aspects of the environment should be included as part of the relevant structure is not to argue that there is selection for environmental properties as well as genotypic properties. That would lead us back to an additive, dichotomous view of development. Instead it is to argue that from a developmental systems perspective there is selection for development relationships (relational properties) not the properties of an entity or object. If the focus in the units of selection debate really is on what counts as an adequate causal account then it is the relationship between the heterozygous genotype and the presence of malaria in the environment that matters. Once again Sober appears to tacitly acknowledge something like this in a footnote. At the bottom of page 48 he reminds the reader to "Note that fitness is properly viewed as a property of the relation of organism to environment, not as an "intrinsic" property of the organism alone".

To summarise the section: I have focused on some of the footnotes in "The Nature of Selection" to foreground certain tensions and contradictions in Sober's text. By highlighting these tensions I have attempted to reveal how Sober's general account of the selection process may be reworked from a contructionist, developmental systems perspective. I have suggested that in Sober's terms, there is *selection of* all the transgenerationally reoccuring components of the developmental system (not just of genes), and *selection for* the properties of developmental relations (not genic, individual or group properties).

4.3 REVERSING THE DISCOURSE

At last we are ready to return to "The Return of the Gene". In this paper Sterelny and Kitcher attempt to defend genic selectionism from the kind of objections raised by Sober. They argue that when biologists such as Dawkins speak of "a gene for that character" they do not really mean that there is a "gene for that character". That would be genetic determinism and preformationism at its worst. Instead, they are just using this phrase as shorthand for the developmentally respectable idea that a genetic difference will cause a difference in an organism's phenotype in a given environment. In philosopher-speak this translates as:

"An allele A at a locus L in a species S is for the trait P* (assumed to be a determinate form of the determinable characteristic P) relative to a local allele B and an environment E just in case (a) L affects the form of P in S, (b) E is a standard environment, and (c) in E organisms that are AB have phenotype P*."[63]

Sterelny and Kitcher claim that this relativising of the effects of an allele to both the allelic and extra-genomic environment means that a coherent genes' eye account can now be given of the cases that Sober argued were problematic for genic selectionism. They subvert Sober's analysis in the following way. Consider (once again) the case of heterozygote advantage. In this case the effects of the "sickling" allele will be dependent on the presence or absence of malaria in the extragenomic environment *and* on the allelic environment. The "sickling" allele is only advantageous in malaria infested regions and when the other allele at the locus is the standard allele. Sterelny and Kitcher conclude that this demonstrates that genic selectionists can indeed give a coherent account of the alleles "having (environment-sensitive) effects, effects in virtue of which they are selected for or selected against." Genic sensitivity to the allelic environment is just a special case of environmental sensitivity. If, in the classic case of industrial melanism, the fact that the fitness of melanic moths varies with the changes in the amount of pollution, is not an argument against the idea that there is *selection for* melanism, then Sober cannot oppose genic selectionism on the grounds that the effects of a gene vary with changes in the allelic context. Genic selectionism rules, OK!.

As the reader may have guessed I derive a rather different moral from Sterelny and Kitcher's genic story. Rather than causing me to conclude that genic selectionism does indeed triumph, I believe that Sterelny and Kitcher's analysis highlights the secondary causal status that both Sober and they accord to aspects of the developmental system that are labelled "environment". For me Sterelny and Kitcher's analysis highlights the fact that Sober treats the external environment in a different way from the allelic context. Combinations of internal factors count as "positive causal factors" (i.e. factors that there is *selection for*), external factors do not. Sterelny and Kitcher's analysis is subversive in that they relabel some aspects of the developmental system (the other allele at the locus) as environment and hence relegate it to a secondary role. From a developmental systems perspective Sterelny and Kitcher's analysis merely highlights how arbitrary the labels we attach to aspects of the developmental system are[64]. Redrawing the line between "internal" and "external" does nothing to change developmental and evolutionary causation, and does not alter the factors that must be jointly specified in an adequate selective explanation. But it is precisely this idea of joint specification and joint or relational properties, that Sterelny and Kitcher's analysis is trying to avoid. They are attempting to reduce the three termed relation between genotype, phenotype, and environment to a two termed relation between genotype and phenotype. Joint properties are thus reduced to properties of individual alleles that are environmentally sensitive.

Alert readers may have noticed an additional way in which Sterelny and Kitcher attempt to diminish the role of the "external" aspects of the developmental system. In their definition of what it means to say that there is a "gene for a character" they relativise the effects of a gene to a "standard environment" rather than just a "given environment". Once again this is clearly a strategy to weaken the causal role of the "environment". If the environment is merely standard background then most of the work can be done by the gene.

The arbitrariness of Sterelny and Kitcher's attempt to privilege the gene as the one aspect of the developmental system that is the real causal agent is best revealed by reversing the story they wish to tell. If "internal" and "external" factors have equal causal status then we should be equally able to speak of "an environment for a character" as we can of "a gene for a character". This would, of course, be shorthand for the idea that an environmental difference would produce a phenotypic difference relative to a given, or standard, genetic background. This reversal of the gene's eye discourse is precisely that strategy that Bateson adopted in his response to "The Selfish Gene".[65] Bateson suggested that if an animal was just a gene's way of making more genes why couldn't we equally say that a bird was just a nest's way of making more nests. Dawkins responded to this suggestion by claiming that the nest could not function as a replicator because variation in nest construction would not be perpetuated in future generations of nests.[66] Clearly, developmental systems theorists would disagree with this claim. Should variations arise in the kind of material that nests are constructed from, then these variations could be "passed on" (=reconstructed) across generations by social traditions. For example, if offspring tend to build nests of similar materials to those they were raised in, then the new nest design would be perpetuated. Variations in nest design could also be perpetuated as a consequence of habitat shifts. If a population shifted into a new environment (e.g., a native New Zealand bird colonised an exotic pine forest), and tended to remain in that habitat as a consequence of habitat imprinting, then the incorporation of new nest material (e.g., pine needles) from the habitat would be reliably perpetuated.

Lots of fun could be had with these environmentalist inversions of the gene's eye view of evolution. For example, instead of the story of the selfish gene imagine the story of the selfish oxygen. In the evolution of the earth's atmosphere oxygen was engaged in intense competition with other atmospheric gases. With the construction of green plants oxygen developed a vehicle for its efficient replication. Chlorophyll containing organisms were thus just oxygen's way of making more oxygen.

Amusing as they might be, these inverted narratives, that privilege the environment rather than the gene, are not the stories I wish to tell. Reversing the discourse is just a tactic to reveal how arbitrary it is to single out one aspect of the developmental system and narrate our tales of development and evolution as if it was the centre of control, and all else was mere background. A constructionist perspective would no more oppose gene centred stories of evolution with environment centred ones, than it would oppose nativistic explanations of development with learning based accounts. In both development and evolution control is not localised in any entity. If information and causation are our focus then it is relationships not entities that count. The basic error of these entity based accounts is that they take networks of co-defining, co-constructing causes and attribute control to just one element in the network. Sterelny and Kitcher attempt to do full justice to the complexities of these causal networks but are constrained by the gene's eye view to always relegate what is labelled "environment" to a secondary role.

Sterelny and Kitcher are not, however, rigid genic selectionists. They advocate what they term pluralistic genic selectionism. That is, they take Dawkins' Necker cube analogy quite seriously. Dawkins argued that just as the Necker cube can be seen from different perspectives, so too can the action of natural selection. Sterelny and Kitcher claim that, "there are alternative, maximally adequate representations of the causal structure of the selection process."[67] For Sterelny and Kitcher the main benefit of a genic representation of selection is its generality. Genic representations of selection are possible when individualistic or group accounts are not. If the only possible forms of representation to choose between are in terms of genes, individuals, or groups then this pragmatic argument is fair enough. However, if a developmental systems account is also possible, then genic representations will not be possible in cases where there is a change in extra-genetic inheritance without a causally related change in gene frequency. Genic representations fail to incorporate all the forms of extragenetic inheritance listed in table 1. To sum up, in response to Sterelny and Kitcher, I would argue not only that a developmental systems account provides a superior representation of the causal structure of the selection process, but also that it is more general.

5. THE EXTENDED PHENOTYPE REVISITED - OF KAKAPO, BUSH FIRES AND SALT-PAN BIOTAS

Ironically, perhaps, the developmental systems perspective shares some features with Dawkins' vision of the extended phenotypes. It shares with Dawkins's vision a denial that the skin is an important dividing line in developmental and evolutionary causation. It shares with Dawkins's vision an emphasis on complex ecological webs of interaction and dependence - extended phenotypes. However, that is where the similarity ends. Dawkins takes these ecological webs (Darwin's "tangled bank") and attributes all the causal power to the genes. From a developmental systems perspective causal power is diffused throughout the web not localised in any single component.

Throughout most of this paper my discussion of the evolution of developmental systems has been fairly abstract. In this section I would like to outline some more concrete examples of the type of evolutionary narratives that arise when evolution is seen from a developmental systems perspective.

So how can we tell the story of the evolution of these ecological webs? Let's start with an example based on one of New Zealand's endemic parrots, the kakapo. The kakapo is a very large and extremely rare flightless parrot (at last count, there were only 49 individuals left). In the breeding season, male kakapo attempt to attract mates using a system of tracks and bowls they have established in their habitat. Tracks are paths the birds have made through the bush. These tracks typically lead up ridges to elevated sites where the males have created bowls (large hollowed-out sites in the ground). To attract females males stand in the middle of these bowls, puff themselves up and emit a powerful booming call. From a developmental systems perspective there are a number of interesting features about the kakapo mating system. First, the birds actively modify their habitat. When a few kakapo were recently transferred to Little Barrier Island (a northern island with no kakapo and no system of tracks and bowls) they reconstructed this feature of their environment (the tracks and bowls) and successfully bred. Second, it seems likely that males who utilise the system of tracks and bowls are at an advantage, as the bowls are generally placed at sites from where the calls can be heard over a wide area (highly advantageous given the low probability of encountering one of the few remaining females). Third, these modifications of the habitat are inherited. That is the systems of tracks and bowls are passed on and maintained across generations. Whereas genic selectionists such as Sterelny and Kitcher would merely regard the tracks and bowls as part of the kakapo's "standard environment", and thus not part of the unit of selection, developmental systems theorists would emphasise the contribution the inheritance of this "standard environment" makes to the kakapo's reproductive success. Kakapo that inherit a well developed track and bowl system will probably have an advantage over those who inherit a poorly developed system (all other factors being equal). This will lead to the differential replication of the kakapo's track and bowl systems, as well as their genes.

This kind of coupled account of the evolution of the landscape and life can be seen on larger scale in the relationship between the evolution of bushfires and the evolution of eucalypts in Australia. Changes in the relationship between a population and its environment (that lead to the differential replication of a developmental system) need not always be initiated by the organisms. The evolution of eucalyptus developmental systems underwent a major transition about 15 million years ago as the lush Australian rainforest dried out. This increasing aridity favoured eucalypts over rainforest vegetation. The features that enabled eucalypts to survive drought, such as their small leaves, thick bark and epicormic shoots than can grow rapidly after drought, also enabled them to survive fires. In fact, the way in which eucalypts regularly shed leaves and bark that are rich in volatile oils, creates an ideal litter environment for the ignition and spread of fire. By modifying their "environment" in this way eucalypts' developmental systems are able to exploit their superior fire tolerance over forms of vegetation that would dominate in the absence of fires. Some eucalypts can exploit the regular occurance of fires as part of their developmental system in an additional way. Their seed capsules (gumnuts) release their seeds after a fire. This enables them to germinate in soil that is not only full of nutrients released by the fire, but also has few other seedlings competing for light and space. This very brief and highly simplified account of the coupled evolution of bush fires and eucalypts could easily be fleshed out to include the implications of these changes for the coevolving biota[68]. The main point I wish to make, however, is that a richer evolutionary story can be told if the focus is on the coupled history of the landscape and its biota - ways in which the landscape changes the biota and the biota changes the landscape - rather than on the evolution of gene pools against an environmental background.

In backgrounding the role of what is labelled "the environment" genic and individualistic selectionists miss out on the possibility of giving an historical explanation for all the transgenerationally stable features of developmental systems. Sterelny and Kitcher, for example, take the "standard environment" as given rather than as a target for additional explanation. Consider the questions a genic selectionist might ask about the following facts. In the inland Central Otago region of New Zealand there are small saline soil areas (salt-pans). Living on these inland salt-pans are a range of "coastal" plants and insects that are either normally found on the coast or are closely related to species found at coastal sites. Faced with these facts the genic selectionist might speculate about the evolution of genes for salt tolerance and how these genes had enabled their phenotypic vehicles to colonise both coastal and inland salt-pan environments. The occurrence of coastal insects on the inland areas could, perhaps, then be explained as a consequence of the plants providing a suitable habitat. In contrast a developmental systems theorist would ask additional questions about the coupled evolution of the life and land. Could it be that both the saline soil and the coastal life are part of a long historical association? Can an historical explanation be given to the whole system rather than just some of its component parts? Obviously these are empirical questions requiring further investigation, but there is indeed some evidence that suggests that the inland saline areas were once coastal sites[69]. With the retreat of the sea the salty soil and coastal life may have remained. If earth and life evolve together in this way then there is more to be explained than the evolution of gene pools.

6. DAWKINSPEAK AND DAWKINSPRACTICE

Sterelny and Kitcher close their paper with an interesting and revealing concession about the dangers of genic selectionism.

"Genic selectionism can easily slide into naive adaptationism as one comes to credit the individual alleles with powers that enable them to operate independently of one another. The move from the "genes for P" locution to the claim that selection can fashion P independently of other traits of the organism is perennially tempting. But, in our version, genic representations must be constructed in full recognition of the possibilities for constraints in gene-environment coevolution. The dangers of genic selectionism, illustrated in some of Dawkins's own writings, are that the commitment to the complexity of the allelic environment is forgotten in practice."[70]

The final brief, but very important argument I wish to raise is a pragmatic one. Our representations of evolution are not neutral with respect to the type of research they encourage. Dawkinspeak leads to Dawkinspractice. If the environment is regarded as merely a standard background upon which genes do their causal work, then of course the "complexity of the allelic environment is forgotten in practice." Researchers who adopt a developmental systems perspective do not need an elaborate system of indexing effects relative to standard environments and they do not need to be eternally vigilant against the dangers of genetic determinism, atomism and adaptationism. From a developmental systems perspective interaction, the developmental integration of networks of mutually contingent causes, and the mutual construction of organism and environment are the primary objects of study rather than secondary features that must be tacked on to make the story work.

7. CONCLUSION

As a parting gesture I would like to explain the rather provocative title of this paper. In announcing "The death of the gene" I do not wish to argue that genes are not important in development. What I wish to lay to rest is the notion of the gene as a "master molecule" controlling and organising development. I wish to dislodge the gene from the privileged site it has occupied in our accounts of development and evolution. This de-centering of the gene within our evolutionary narratives should not be seen as an argument for the return of the organism, the cytoplasm, the environment, or any other developmental resource. Instead, in this paper I have aimed to illustrate how evolution can be narrated from a developmental systems perspective that does not privilege any component of the system. This account is very preliminary. However, I believe that it has suggested that there might be a far more interesting story to be told about evolution than the units of selection controversy has suggested to date. It is a richer story about the evolution of developmental systems, of extended phenotypes, and of the coupled differentiation of life and land. But it is a story that will not be developed while we continue to remain trapped, endlessly debating whether the unit of selection is the individual or the gene. There are more than two stories to be told and, as Haraway notes, "In more ways than one, one story is not as good as another."[71]

ACKNOWLEDGEMENTS

This paper owes a great debt to Pat Bateson, Gilbert Gottlieb, Timothy Johnston, Richard Lewontin and Susan Oyama. Their papers, correspondence and conversations provided the sources for many of the ideas I have expressed here. Nicola Gavey, Martyn Kennedy and Myrna Wooding all assisted in the production of this paper. Finally I would like to thank Paul Griffiths for his ideas and encouragement in spite of the fact I must have strained his editorial patience to its absolute limit.

NOTES

[*] Adapted from Bateson, P. (1983) 'Genes, Environment and the Development of Behaviour', in *Animal Behaviour: Genes, Development and Learning* , Slater, P. & Halliday, T. (eds) Blackwell, by permission of the publishers.

[**] Adapted from Clausen, J., Keck, D.D., & Hiesey, W. M. (1948) *Experimental Studies in the Nature of Species Vol. 3, Environmental Responses of Climatic Races of Achillea* Carnegie Institute of Washington Publications, No. 581, p80, by permission of the publishers.

[1] Haraway, D. (1989) *Primate Visions: Gender, Race and Nature in the World of Modern Science,* Routledge. p8.

[2] Sterelny, K. & Kitcher, P. (1988) 'The Return of the Gene', *The Journal of Philosophy* **LXXXV**, p339-361.

[3] This debate has produced some truly wonderful papers and exchanges - see Kuo, Z.-Y. (1921) 'Giving up Instincts in Psychology'. *Journal of Philosophy* **XVIII**, p645-664; Kuo, Z.-Y. (1976) *The Dynamics*

of Behavior Development (enlarged ed.), Random House; Hebb, D.O. (1953) 'Heredity and Environment in Animal Behaviour', *British Journal of Animal Behaviour* **1**, p43-47; Lehrman, D.S. (1953) 'A Critique of Konrad Lorenz's Theory of Instinctive Behaviour', *The Quarterly Review of Biology* **28**, p337-363; Lehrman, D.S. (1970) 'Semantic and Conceptual Issues in the Nature-Nurture Problem', in *Development and Evolution of Behavior*, Aronson, L.R., Tobach, E., Lehrman, D.S., & Rosenblatt, J.S. (eds.), Freeman; Schneirla, T.C. (1956) 'Interrelationships of the "Innate" and the "Acquired" in Instinctive Behavior', in *L'Instinct dans le Comportement des Animaux et de L'Homme*, P.-P. Grasse (ed.), Masson, p387-452; Hinde, R.A. (1968) 'Dichotomies in the Study of Development', in *Genetic and Environmental Influences on Behaviour*, Thoday, J.M. & Parkes, A.S. (eds.), Oliver & Boyd, p3-14; Gottlieb, G. (1976) 'Conceptions of Prenatal Development: Behavioral Embryology', *Psychological Review* **83**, p215-234; Bateson, P. (1983) 'Genes, Environment and the Development of Behaviour' in *Animal Behaviour: Genes, Development and Learning*, Slater, P. & Halliday, T. (eds.), Blackwell, p52-81; Oyama, S. (1985) *The Ontogeny of Information : Developmental Systems and Evolution*, Cambridge University Press; Johnston, T.D. (1987). 'The Persistance of Dichotomies in the Study of Behavioral Development', *Developmental Review* **7**, p149-182.

[4] Most, but not all. See Johnston, T.D. (1988) 'Developmental explanation and the Ontogeny of Birdsong: Nature-Nurture redux', *Behavioral and Brain Sciences* **11**, p617-663 for a critique of the recent view that there are innate templates underlying the development of birdsong.

[5] This list of objectives is based on the arguments of Kuo, Hebb, Lehrman, Schneirla, Hinde, Gottlieb, Bateson, Oyama and Johnston in the papers listed above.

[6] Bateson, P. (1991). 'Are there Principles of Behavioural Development', in *The Development and Integration of Behaviour*, Bateson, P. (ed) Cambridge University Press, p21.

[7] Hailman, J.P. (1967) 'The Ontogeny of an Instinct. The Pecking response in Chicks of the Laughing Gull (Larus atricilla L.) and Related

Species', *Behaviour Supplement* **15**, p1-159.

[8] e.g., Lehrman (1970) op. cit.

[9] See Lehrman (1970) op. cit., Bateson (1983) op. cit, and Johnston (1987) op. cit.

[10] See Bateson, P.P.G. (1976). 'Specificity and the Origins of Behaviour', *Advances in the Study of Behavior* **6**, p1-20, and Bateson (1983) op. cit.

[11] For interesting discussions of the problems of genetic determinism and the complexity of developmental processes at the molecular level see Stent, G. (1981) 'Strength and Weakness of the Genetic Approach to the Development of the Nervous System' in *Studies in Developmental Neurobiology*, Cowan, W.M. (ed.), Oxford University Press; Tapper, R. (1989) 'Changing messages in the Genes,' *New Scientist* **25 March**, p53-55; Nijhout, H.F. (1990) 'Metaphors and the Role of Genes in Development', *BioEssays* **12**, p441-446.

[12] See Lickliter, R. & Berry, T.D. (1990) 'The Phylogeny Fallacy: Developmental Psychology's Misapplication of Evolutionary Theory,' *Developmental Review* **10**, p348-364.

[13] See Gottlieb, G. (1981) 'Roles of Early Experience in Species-specific Perceptual Development' in *Development of Perception*, **1**, Aslin, R.N., Alberts, J.R., & Petersen, M.P. (eds.), Academic Press.

[14] For a good discussion of the importance of these general experiential inputs into development see Bateson (1976) op. cit.

[15] Dudai, Y., Jan, Y. -N., Byers, D., Quinn, W.G. & Benzer, S. (1976) '*dunce*, a mutant of *Drosophils* Deficient in Learning', *Proceedings of the National Academy of Sciences, USA* **73**, p1684-1688.

[16] Sackett, G.P., Ruppenthal, G.C., Fahrenbruch, C.E. & Holm, R.A. (1981) Social Isolation Rearing Effects in Monkeys vary with Genotype, *Development Psychology* **17**, p313-318.

[17] See Oyama (1985), op. cit, and Johnston (1987), op. cit.

[18] A major section of a recent issue of the journal Bioscience was devoted to norms of reaction. *Bioscience* 1989, **39**.

[19] Lewontin, R.C. (1982) Human Diversity. *Scientific American*, p21.

[20] Lewontin (1982), op. cit, p22.

[21] Ho, M.-W. (1986) 'Heredity as Process: Towards a Radical Reformulation of Heredity', *Rivista di Biologia/Biology Forum* **79**, p407-447.

[22] See Lewontin, R.C. (1983) 'The Organism as the Subject and Object of Evolution,' *Scientia* **118**, p65-82, and Gray, R.D. (1988) 'Metaphors and Methods: Behavioural Ecology, Panbiogeography and the Evolving Synthesis,' in *Evolutionary Processes and Metaphors*, Ho, M.-W. & Fox, S.W. (eds.), Wiley, for a discussion of the importance of these modifications.

[23] Although this view is really quite different from past perspectives, it obviously builds on the insights of the past.

[24] I have taken this label from the work of Lewontin and Oyama. A list of papers that present a constructionist view might include Stent (1981) op. cit; Lewontin (1983) op. cit; Oyama (1985) op. cit; Oyama, S. (1988) 'Stasis, Development and Heredity' in *Evolutionary Processes and Metaphor*, Ho, M.-W. & Fox, S. W. (eds.), Wiley; Oyama, S. (1989) 'Ontogeny and the Central Dogma: Do we Need the Concept of Genetic Programming in Order to have an Evolutionary Perspective?', in *Systems and Development. The Minnesota Symposia on Child Psychology*, **22**, Gunnar, M. R. & Thelen, E. (eds.), Erlbaum; Oyama, S. (in press) 'Transmission and Construction: Levels and the Problem of Heredity', in *Critical Analyses of Evolutionary Theories of Social Behavior: Genetics and Levels*, Greenberg, G. & Tobach, E. (eds.), Shapolsky; Ho (1986) op. cit; Ho (1988); Gray, R. D. (1987). 'Faith and foraging: A Critique of the "Paradigm Argument from Designs"' in *Foraging*

Behavior, Kamil, A. C., Krebs, J.R. & Pulliam, H.R. (eds.), Plenum; Johnston, T. & Gottlieb, G. (1990) 'Neophenogenesis: A Developmental Theory of Phenotypic Evolution,' *Journal of Theoretical Biology* **147**, p471-495; Lickliter & Berry (1990) op. cit.

[25] Richard Dawkins, who is not always the crude genetic determinist his critics sometimes claim, notes something like this. "Genetic causes and environmental causes are in principle no different from each other," Dawkins, R. (1982) *The Extended Phenotype: The Gene as the Unit of Selection*, p 13, Freeman. It is worth pointing out, however, that the phrase developmental context includes a lot more factors than just the "environment" (e.g., cytoplasmic factors, self-stimulation, etc.).

[26] Stent (1981) op. cit, Lewontin (1982) op. cit, Nijhout (1990) op. cit.

[27] Lewontin (1982) op. cit. This example is based on the work of Conrad Waddington. See Waddington, C.H. (1975) *The Evolution of an Evolutionist*, Cornell University Press.

[28] This is, of course, just one example of the general phenomena of sensitive periods in development. For a review see Bateson, P.P.G. (1979). 'How do sensitive periods arise and what are they for?', *Animal Behaviour* **27**, p470-486.

[29] See Shaw, R. & Turvey, M. T. (1991) 'Coalitions as models for ecosystems: a realist perspective on perceptual organisation' in *Perceptual Organisation*, Kubany, M. & Pomerantz, J.R. (eds.), Erlbaum; Lewontin (1983), op. cit, and Gray (1988), op. cit for a more detailed account of this.

[30] Oyama (1985) op. cit.

[31] This figure is adapted from Wilden, A. (1980) *System and Structure: Essays in Communication and Exchange*, 2nd ed., Tavistock, by permission of the publishers.

[32] The importance of this expanded view of inheritance is emphasised

by Oyama (1985) and (1989) op. cit, Gray (1987) and (1988) op. cit and Socha, R. (1990) 'Beyond Genocentric Concept of Heredity and Evolution' in *Evolutionary Biology: Theory and Practice,* **Part II.** Leonovicova, V.V., Novak, V.J.A., Slipha, J. & Zemek, K. (eds.), Czechoslovak Academy of Sciences.

[33] Cohen, J. (1979) 'Maternal Constraints on Development', *Maternal Effects in Development,* Newth, D.R. & Balls, M. (eds.), Cambridge University Press; Ho, M.-W. (1984) 'Environment and Heredity in Development and Evolution', *Beyond Neo-Darwinism: An Introduction to the New Evolutionary Paradigm,* Ho, M.-W. & Saunders, P.T. (eds.), Academic Press; Galef, B.G.(Jr) & Henderson, P.W. (1972) 'Mother's Milk: A Determinant of the Feeding Preferences of Weaning Rat Pups', *Journal Comparative and Physiological Psychology* **78**, p213-219; Corbet, S.A. (1985) 'Insect Chemosensory Responses: A Chemical Legacy Hypothesis', *Ecological Entomology* **10**, p147-153; Hepper, P.G. (1988) 'Adaptive Foetal Learning: Prenatal Exposure to Garlic Affects Postnatal Preferences', *Animal Behaviour* **36**, p935-936; Mattson, W.J.(Jr) (1980) 'Herbivory in Relation to Plant, Nitrogen Content', *Annual Review Ecology Systematics* **11**, p119-161; Jones, G.R., Aldrich, J.R. & Blum, M.S. (1981) 'Baldcypress Allelochemicals and the Inhibition of Silkworm Enteric Microorganisms, Some Ecological Considerations', *Journal of Chemical Ecology* **7**, p103-114; Fisher, J. & Hinde, R.A. (1949) 'The Opening of Milk Bottles by Birds', *British Birds* **42**, p347-357; Norton-Griffiths, M. (1968) *The Feeding Behaviour of the Oystercatcher (Haemotopus ostralegus),* Ph.D. Thesis, Oxford University; Van Denburgh, J. (1914) 'The Gigantic Land Tortoises of the Galapagos Archipelago', *Proceedings of the Californian Academy of Sciences,* **San Francisco 4th ser. 2**, p203-374; Helfman, G.S. & Schultz, E.T. (1984) 'Social Transmission of Behavioural Traditions in a Coral Reef Fish', *Animal Behaviour* **32**, p 379-384; Neal, E. (1948) *The Badger,* Collins; Carrick, R. (1963) 'Ecological Significance of Territory in the Australian Magpie, Gynorhina tibicen', *Proceedings of the International Ornithological Congress,* **2**, p740-753; Jolly, A. (1972) *The Evolution of Primate Behaviour,* Macmillan; Woolfenden, G.E. & Fitzpatrick, J.W. (1978) 'The Inheritance of Territory in Group Breeding Birds', *Bioscience* **28**, p104-108; Harris, M.A. & Murie, J.O. (1984) 'Inheritance of Nest Sites in Female Columbian Ground Squirrels',

Behavioral Ecology Sociobiology **15**, p97-102; Cheney, D.L. (1977) 'The Acquisition of Rank and the Development of Reciprocal Alliances Among Free-Ranging Immature Baboons', *Behavioral Ecology Sociobiology* **2**, p303-318; Harrocks, J. & Hunte, W. (1983) 'Maternal Rank and Offspring Rank in Vervet Monkeys: An Appraisal of the Mechanisms of Rank Acquisition', *Animal Behaviour* **31**, p772-782; Jenkins, P.F. (1978) 'Cultural Transmission of Song Patterns and Dialect Development in a Free-Living Bird Population', *Animal Behaviour* **25**, p50-78; Croizat, L. (1964) *Space, Time, Forum: The Biological Synthesis*, published by the author; Rosen, D.E. (1978) 'Vicariant Patterns and Historical Explanation in Biogeography', *Systematic Zoology* **27**, p1-16., Jablonski, D. (1987) Heritability at species level: Analysis of Geographic ranges of Cretaceous Molluscs, *Science* **238**, p360-363.

[34] See Vygotsky, L. (1978) *Mind in Society*, Harvard University Press; Kaye, K. (1982) *The Mental and Social Life of Babies*, University of Chicago Press; and Valsiner, J. (1987) *Culture and the Development of Children's Action*, Wiley.

[35] The extent to which earth and life function as a co-evolutionary unit is currently a hot topic in the field of historical biogeography. For a New Zealand perspective on the debate see the special issue of the *New Zealand Journal of Zoology* on panbiogeography, **16** (1989).

[36] Compare with Boyd, R. & Richerson, P.J. (1985). *Culture and the Evolutionary Process*, University of Chicago Press; and Odling-Smee, F.J. (1988) 'Niche-Constructing Phenotypes', in *The Role of Behavior in Evolution*, Plotkin, H.C. (ed.), MIT Press.

[37] This cake analogy has been used by Bateson (1976) op. cit and Dawkins, R. (1981) 'In Defense of Selfish Genes', *Philosophy* **56**, p556-573 to emphasise that there is no one-to-one correspondence between developmental ingredients and the phenotypic cake. The cake analogy, while often useful, also has its limitations. Once again the ghost of pre-existing design could rear its head. There are recipes for cakes, just as it is claimed that there are genes for phenotypes. If development is like baking a cake, then it is in an unusual kitchen where there are no

recipes and the cake constructs itself (i.e. the cake is self-organising).

[38] Stent (1981) op. cit.

[39] Oyama (1985) op. cit and (1989) op. cit.

[40] See Dobzhansky, T. (1951) *Genetics and the Origin of Species*, 3rd ed., Columbia University Press, p16 and Dawkins, R. (1976) The Selfish Gene, Oxford University Press, p48.

[41] See Oyama (1985) op. cit, Oyama (1989) op. cit, Gray (1987) op. cit, Gray (1988) op. cit, Johnston and Gottlieb (1990) op. cit, Lichliter and Berry (1990) op. cit, and Socha (1990) op. cit.

[42] See Costall, A. (1986) 'Evolutionary Gradualism and the Study of Development'. *Human Development* **29**, p4-11.

[43] The phenotype here is an extended phenotype that includes relevant features of the environment.

[44] Lewontin, R.C. (1974) *The Genetic Basis of Evolutionary Change*, Columbia University Press, p19.

[45] See Johnston and Gottlieb (1990) op. cit. for a more detailed argument.

[46] Dawkins, R. (1976) *The Selfish Gene*, Oxford University Press.

[47] See Hamilton, W.D. (1964) 'The Genetic Theory of Social Behavior', I & II, *Journal of Theoretical Biology* **7**, p1-32; Williams, G.C. (1966) *Adaptation and Natural Selection*, Princeton University Press.

[48] Dawkins (1976) op. cit, p21.

[49] Dawkins, R. (1982). *The Extended Phenotype: The Gene as the Unit of Selection*, Oxford University Press.

[50] Dawkins, R. (1984) "Replicator Selection and the Extended Pheno-

type' in *Conceptual Issues in Evolutionary Biology*, Sober, E. (ed.), MIT Press, p138.

[51] Dawkins (1984) op. cit, p139.

[52] Gould, S.J. (1980) *The Panda's Thumb*, W.W. Norton.

[53] Bateson, P.P.G. (1978) 'Book Review: The Selfish Gene'. *Animal Behaviour* **26**, p316-318.

[54] Dawkins (1982) op. cit, p 89-99.

[55] Hull, D.L. (1988) *Science as a Process: An Evolutionary Account of the Social and Conceptual Development of Science*, University of Chicago Press, p413.

[56] See Williams, G.C. (1986) 'Comments by George C. Williams on Sober's The Nature of Selection', *Biology and Philosophy* **1**, p114-122 and Sober, E. (1984) *The Nature of Selection: Evolutionary Theory in Philosophical Focus*, MIT Press.

[57] Williams (1986) op. cit, p121.

[58] Hull (1988) op. cit, p409.

[59] Sober (1984) op. cit.

[60] Sober, E. (1984) 'Force and Disposition in Evolutionary Theory' in *Minds, Machines and Evolution*, Hooking, C. (ed.), Cambridge University Press, p50-51.

[61] In adopting Sober's terminology I am not necessarily endorsing his more general views on causation.

[62] Lloyd, E.A. (1988) *The Structure and Confirmation of Evolutionary Theory*, Greenwood Press, p136.

[63] Sterelny & Kitcher (1988) op. cit, p350.

[64] Lloyd (1988) op. cit, p350.

[65] Bateson (1978) op. cit.

[66] Dawkins (1984) op. cit.

[67] Sterelny & Kitcher (1988) op. cit, p358.

[68] For more on eucalypts and bushfires see Pryer, L.D. (1976) *Biology of Eucalypts*, Edward Arnold; and Pate, J.S. & McComb, A.J. (1981) *The Biology of Austraiian Plants*, University of Western Australia Press.

[69] Patrick, B. (1989), 'Panbiogeography and the Amateur Naturalist with reference to Conservation Implication', *New Zealand Journal of Zoology* **16**, p749-755.

[70] Sterelny & Kitcher (1988) op.cit. p360-361.

[71] Haraway (1989) op. cit, p358.

Russell Gray,
Department of Psychology,
University of Otago.

Susan Oyama

ONTOGENY AND PHYLOGENY; A CASE OF METARECAPITULATION?

For some time I have been occupied with the nature-nurture opposition (genes-environment, innate-acquired, etc.)[1] Over this period I have become sensitised, not only to the various guises in which this dichotomy appears, but also to structurally similar ones in other fields. (Fig. 1) In epistemology, one of the sources of the nature-nurture dichotomy in science, the classical question has concerned the origin of knowledge. The disputes between rationalists, who insisted upon innate knowledge, and empiricists, who credited the senses, did much to set the framework for more recent disputes. Similar oppositions are found in other fields.

SELECTED DICHOTOMIES

Epistemology	:	innate vs experiential sources of knowledge.
History	:	internalist vs externalist explanations of change.
Anthropology	:	biology vs culture. cultural vs geographical determinism.
Biology	:	history vs ecology. physical necessity vs chance in explaining the origins of life. mosaic vs regulative development in embryology.

Figure 1. Dichotomies in which internal factors are opposed to external ones.

P. Griffiths (ed.), Trees of Life, 211–239.

Whereas psychology, biology, and contemporary philosophy inherited the epistemological question fairly directly, a number of other contrasting pairs can be found that have less obvious ties to the nature-nurture complex per se. Among historians, one finds internalist and externalist views of historical change, including scientific change.

Anthropologists attribute cultural patterns in a society to biology or to culture (a form of nature-nurture dichotomy), and, interestingly enough, argue about cultural versus geographical determinism as well.[2]

In contrasting inherited ways of life with immediate conditions of life, the last example resembles the history-ecology distinction in biology: Russell Gray has discussed ecologists' pitting of historical factors against ecological ones in explaining relationships between populations and their niches. Other biological oppositions can also be found. Students of the origin of life emphasise either physical necessity or chance. Embryology, meanwhile, has had its disputes over mosaic and regulative development.[3]

Not all these dichotomies are equally similar: tracing their histories and relations would be a fascinating exercise in itself. I would like, instead, to examine some striking resemblances between the nature-nurture dispute in developmental studies and the argument over developmental constraints and natural selection as competing explanations of evolution. My suggestion is that the sorts of conceptual problems that are generated in the first may also arise in the second (hence the "metarecapitulation" of my title). In past work, I have elaborated a view of development that eliminates the need for the developmental dualism of the nature-nurture opposition; here I sketch a related formulation of evolutionary change that may eliminate the opposition between constraint and selection as well.

Adequate definitions, consistency, and conceptual clarity are, in my view, reason enough for the kind of analysis engaged in here. There are other reasons as well. In the nature-nurture debate, conceptual problems have had significant consequences.

It is true that when ambiguous terminology leads to misunderstandings, even serious ones, the remedy appears to be straightforward: be a bit more careful with your words. Indeed, conceptual analysis is often dismissed as mere quibbling over what things are called. Usually, however, matters are not so simple. When ambiguities and conflations are persistent and recurrent, as they are in the nature-nurture opposition, it is because there are serious difficulties with the reasoning itself. Faulty inference and outright incoherence remain common even after superficial adjustments in lexicon are made. Because of the social, political, and moral implications of the various notions of biological nature, furthermore, these problems are not confined to scholarly matters, but become entwined in larger debates about individual and group differences, the necessity of particular social arrangements, and so on. Conceptual realignment is thus required; simply changing a few labels is not enough.

In addition, I am convinced that it is useful to examine the often subtle assumptions embedded in theory. These assumptions about the nature of causality, agency, and process are not only of importance for scientists, but they are also intimately related to our views of our own actions and possibilities.

The question mark in my title is a real one. While I have devoted more than a decade to documenting the mischief made in and by the nature-nurture dichotomy, it is less than obvious that the constraints-selection distinction necessarily involves difficulties of the same variety and magnitude. This essay is an invitation to biologists and philosophers to consider just how far the parallels between the two disputes extend. The disanalogies may be as significant as the analogies. Disputes over evolutionary dynamics, unlike arguments about innate selfishness or aggressiveness, seem removed from everyday matters, and do not appear to have the same sorts of complex implications. Still, as I point out later, the two discourses are not independent.

PARALLELS

An important question in developmental studies, if not the important question, pertains to the ontogenetic origin of organismic form, including the form of the mind.

Traditionally, answers to this question have focused either on a set of basic structures supposed to be "transmitted" in the genetic material or on the contingencies of individual experience.[4] Thus, nature-nurture battles are ostensibly about the allocation of Causal responsibility for development either to the genes or to the environment. The motivating concern, however, often appears to be with limits to phenotypic variation and change.

In evolutionary theory, the prime question is also about the origin of form, but over phylogenetic, not ontogenetic, time: Internal developmental constraints are set against the contingencies of natural selection. Again the questions are couched in terms of the relative influence of alternative causes, and again the basic concern seems to be with limits, but this time the limits on phylogenetic variability and change. In both fields, one finds a kind of causal dualism, as internal forces are opposed to external ones. I submit that these quarrels over causal responsibility miss the point, and that the point in each case is developmental dynamics: the possibilities for alternative developmental outcomes in a single lifetime and the possibilities for change in developmental systems across generations.

A somewhat more detailed account of these parallels follows, but there is also a more direct relationship between the two dichotomies: Ontogeny appears in the evolutionary debate as a constraint on evolutionary change, and the internal, genetic, factors in ontogeny are the legacy of that evolutionary change. (Fig. 2) Since the opposition between nature and nurture is so intimately related to the one between constraints and selection, it seems natural to wonder whether their

	Internal	**External**
Phylogeny	nature	nuture
Ontogeny	constraints	selection

Figure 2. Relationship between dichotomies in development and evolution

resolutions may be related as well.

Of the many points that could be made here I wish to highlight just four.[5] In both debates, internal causes are contrasted with external ones. In both fixity is associated with the former and malleability with the latter, and in both there has been considerable oscillation between internalist and externalist perspectives. Finally, similar compromises have been proposed, as theorists attempt to reconcile internal to external factors. These are considered in turn.

INSIDES AND OUTSIDES

The nature-nurture debate turns on a radical separation of insides from outsides. So does the one about constraints and selection. Richard Lewontin and Elliott Sober have elaborated two models of change that give us a convenient way of seeing how this occurs in each case.[6] One is the transformational model. It can be characterised by two propositions: (1) Change in a collection results from change in the constituent entities and (2) The entities change in a uniform and predetermined manner. In the variational model (which includes selectional processes): (1) Variant entities are propagated with differing frequencies and (2) The variants themselves are static. Change is generated from within in the transformational model and imposed from without in the variational model. (Figure 3).

In developmental theory, the transformational model dominates. Maturation, as it is traditionally conceived, is the quintessential internally driven process.[7] It is to capture the regularity of such ontogenetic sequences that metaphors like the genetic programme are invoked: Biological "nature" is treated as both the cause and the effect of these regular processes. The "programme" that begins by describing the predictability of development thus becomes a quasi-explanatory device, as the reliable reappearance of certain features in successive generations is attributed to a central, internal control. Species-typical processes and characters[8] then constitute the fixed core that is "transmitted" in the genetic material.

The dominance of the transformational model of development is such that even the most committed environmentalist accepts, as a

Transformational :	1. Change in a collection results from change in the constituent entities. 2. The entities change in a uniform and predetermined manner.
Variational :	1. Varient entities are propagated with differing frequencies. 2. The varients themselves are static.

Figure 3. Two Models of Change

matter of course, a genetically given body and a set of reflexes, instincts, or some other substrate for behavioural development. An inherited structure, with all the preformationist assumptions this entails, is a prerequisite for all nature-nurture haggling. Disagreements about whether some particular trait has an ontogenetic or a phylogenetic origin, then, occur against a consensual background of intrinsically driven processes; indeed, developmental studies are virtually defined by the transformational model.[9]

"Nurture" in these exchanges over developmental causation is not so unitary. Selection, a variational process, is frequently contrasted with instruction, for instance. Both are secondary elaborations of a pregiven "biological base", however, and both are seen as externally directed. Selectionist explanations are found in the literatures of operant conditioning, neurogenesis, the immune system, and cognitive and linguistic development.[10]

If the internalist perspective defines developmental studies, the field of evolution presents a rather different picture. Variational processes, including genetic drift and natural selection, dominate the synthetic theory. Populations are "moulded" by these external forces.

Orthogenesis, resoundingly rejected by the modern synthesis, represented a transformational view of evolution. The disappearance of such an internalist, "developmental" perspective from evolutionary theory is generally seen as the triumph of scientific Darwinism over mystical Lamarckian progressivism.[11] Virtually abolished from contemporary evolutionary discourse,[12] "orthogenesis" is probably too heavily freighted a term to gain serious attention now. Its internalist connotations persist, however, in the literature on developmental constraints, a literature that is gaining in volume and influence. A notable attempt to integrate its diverse critical and empirical viewpoints into mainstream neoDarwinism is seen in the group report on a conference convened for just that purpose.[13] In addition to presenting a typology of constraints, the paper repeatedly points out the importance of, and difficulty of, separating internal factors from external ones.

MALLEABILITY AND FIXITY

Variational processes are secondary to transformational ones in conceptualising ontogeny, then, while the reverse is true in conventional thinking about phylogeny: Natural selection tends to be seen as the major formgiver, while developmental constraints serve merely to narrow its scope. This separation of insides from outsides brings us to our second parallel: In both areas, malleability is associated with external factors, and fixity, with internal ones. One even finds the same vocabulary used to express these relations. "Shaping" is frequently seen by both developmentalists and evolutionists as capriciously variable; it represents the vagaries of individual or species history. Shaping and moulding, in fact, are probably the most common metaphors for natural selection, invoking the image of an omnipotent, omnipresent hand and eye (and will), the coordinated apparatus of the artisan. The shaping metaphor obscures the difference between the origin of variants and their perpetuation, a difference emphasised by many developmental constraints theorists.

"Shaping" is also a technical term in operant theory, referring to the gradual production of a complex behaviour pattern that does not occur spontaneously. Reinforcement of successive approximations of that pattern, often by a trainer who progressively raises the performance standards, results in the appearance of the target behaviour. In shaping, then, the problem and the criteria for reinforcement are set by the environment, though not necessarily by a human trainer.[14]

In both cases an emphasis on external shaping involves a

tendency to take the generation of the selected variants for granted, to view it as unproblematic: unsystematic, random, at least with respect to the selection. In the same way that the context of scientific discovery is considered to be independent of, and irrelevant to, the real business of justification, the mechanics of the generation of variation are treated as irrelevant to natural selection, as long as there is heritability (like begets like). To focus on constraints, however, is to ask why some variants arise and not others, and this question tends to be conceived as involving internal causes.

Thus one hears of genetic constraints on learning (species differences, often invoked to explain failure of conditioning), or of the genes limiting possible social arrangements.[15]

Although any genotype is the result of an evolutionary history, in developmental studies it frequently plays the role of a fixed, eternal essence, insulated from change, the unmoved mover that both embodies the plan of an organism and executes it. In similar manner, developmental constraints on natural selection are sometimes presented as timeless physical laws or ahistorical necessity.[16] They limit the otherwise untrammeled variation stemming from environmental fluctuations, though a distinction may be made between universal constraints and those affecting only some evolutionary lines.[17] When constraints are local or historical, they are a genealogical legacy. They carry the lineage history and influence its further history; they are species "nature" to evolutionary "nurture". In either case, developmental constraints tend to be conceptualised as placing limits on the arbitrary action of selection, forbidding certain forms and permitting others.

Resistance to change, then, is more salient in both these literatures than the direction or generation of change, though these are not opposed but are simply two aspects of biological dynamics.[18] The language of direction and guidance can be used to express the same phenomena, of course: Behaviour theorists speak of predispositions to learn and constraints theorists refer to biases and propensities. This dual role of constraints (resisting and directing) does not go unnoticed.[19] To emphasise the orderly generation of variants is to give internal factors a more prominent, creative role, while external factors may then be cast as secondary filters. A constraint, as Stephen Stearns notes, is usually "imported from outside the local context to explain the limits on the patterns observed", and Stephen Gould observes that the term is a theory laden one that indicates the causes being identified are other than the "canonical causes" in the theory.[20]

Though Gould makes a case for retaining the term and enlarging its meaning, it is not surprising that "constraint" tends to be supplanted by other terms when the researchers are making a bid to move their tradition from the periphery to the centre. Grehan and Ainsworth, for instance, seem to take issue with the tendency of the constraints literature to emphasise limitation; their position is that selection is "subsidiary" to orthogenetic tendencies to vary in certain directions. And Goodwin urges a "rewriting of the origin of species so that 'origin' is understood primarily in its logical, generative sense, and secondarily in historical terms".[21]

The language of internal resistance to change (and of internal tendencies channelling change) bring to mind the image of recalcitrant material in the hands of an artisan; it recalls an ancient distinction between matter and form. The artist's freedom is limited by the nature of the material, and the notion of raw material with its own stubbornly causal properties is present in these contemporary debates about ontogenetic and phylogenetic change.[22] Because fixity is attributed to internal causes, and malleability, to external ones, queries that are unintelligible if taken literally (whether internal or external factors are responsible for some feature) may be seen as (admittedly odd) ways of asking how amenable an organism is to various developmental influences, or how susceptible a species' developmental processes are to various kinds of transgenerational change. Living beings, however, are not made of static stuff, so it is dynamic stability that must be explained, and such stability is attained and maintained by interaction, not isolation.

OSCILLATION

The third parallel between the developmental and evolutionary debates is that both have been characterised by an oscillation of received wisdom between the internal and external poles. Early psychology's preoccupation with internal entities like mind, drives, and instincts led to the behaviourist obsession with external causation. This in turn led to the present resurgence of "biological" approaches (behaviour genetics, physiological and evolutionary studies), typically understood as focusing on "intrinsic factors". (Notice again the associations among biology, internality and fixity.)[23] Even more recently, certain fields have witnessed a partial retreat from strict "programming" accounts: The above mentioned selectional theories of neurogenesis, cognition and the immune system are examples. These theories tend to be linked to very traditional maturational stories; selectional theories of language acquisition and cognitive development are, in fact, strongly nativist, both in intellectual lineage and in current orientation.

My earlier point about the preeminence of transformational explanations in developmental studies is relevant here; these research trends do not signal a different way of conceptualising development in general, but rather, they are sophisticated efforts to reconcile traditional maturational explanations with certain empirical challenges: the astonishingly specific responsiveness of the immune system, for instance, or the obvious necessity for a theory of language acquisition to accommodate differences among languages.

In evolutionary theory, the current interest in internal constraints seems largely to be a reaction against the hegemony of natural selectionist explanations, while, as already noted, neo Darwinist selectionism is often seen as the definitive refutation of earlier, more transformational, visions of species change. Predictably, perhaps, some sociobiologists, whose power to explain the world depends heavily on the "power" of natural selection (an idiom we shall return to), have responded to the constraints literature by reaffirming their faith in the omnipotence of selection to mould bodies and minds.[24]

COMPROMISES

Not only does one find oscillation in both developmental and evolutionary debates, but the same (largely unsatisfactory) compromises are encountered as well.[25] Developmentalists have often "solved" the nature-nurture problem by dividing the territory: by attributing some features to genetic control and others to environmental influences. But if there are fundamental conceptual difficulties with the very notion of genetically or environmentally directed development, it can scarcely be an improvement to combine them in some odd organic patchwork.

Apparently more judicious is the attempt to quantify the relative amounts of genetic and environmental influence on various features. There have been many important critiques of these strategies of phenotype partitioning, as well as of the partitioning of variance associated with statistical techniques like the analysis of variance (ANOVA); even the seemingly more sophisticated tack of tracing phenotypic "information" to the genes and the environment is problematic.[26] One difficulty with the partitioning of variance is that it frequently slides into partitioning phenotypes. Another is that, even if the eye is fixed firmly on variance, not on variants, the results of local analyses tend to be confused with general functional relationships.[27] A third, related, one is that the ambiguous use of terms like "genetic control" both to explain invariance (inevitability or species typicality, for instance) and variance (genotype associated variation in ANOVA) encourages exactly the erroneous inference of developmental fixity (of IQ, for instance) from heritability coefficients that has plagued behaviour genetics for decades.

Largely in an attempt to avoid such interpretational difficulties, some theorists have tried another tack: genetic imperialism, in which genes are said to determine the range of possibilities, while the environment merely selects the particular value. The genes are given higher-level, generalised control, while secondary influence is doled out to non-genetic factors.[28] The genes cannot, of course, define an array of possible outcomes independently of an array of developmental environments (each of which is actually an indefinitely long sequence of nested environments at various scales). There is a certain vacuity, furthermore, to declaring that only those phenotypes that can develop will develop, and it certainly makes no sense to place determinative power in only one set of interactants when every developmental outcome is jointly specified by genotype-environment (actually, organism-environment) pairings.

It is not accidental that both partitioning and imperialistic conquest are typical outcomes of geopolitical conflict. Nor do I use these terms unreflectively. Indeed, I see nature-nurture arguments as territorial disputes in which the contenders strive to retain as much power (to explain, to control what counts as data, to dominate a field of inquiry, to subsume related areas) as they can, even as they make necessary concessions. Nor do I think issues of "turf" are trivial or devoid of intellectual significance, though they are often presented that way. They represent entire traditions of theory and research, which in turn represent conceptions of science, of scientists, and of the world.[29]

As the reader has no doubt realised, these styles of compromise appear in the evolutionary literature as well. When theorists attribute a character to selection or to constraints, they are implying that species features can be credited to alternative formative factors. Asking how much a feature owes to each factor, meanwhile, is analogous to the quantifying of genetic and environmental influences on a phenotypic character.[30] When, in commenting on this literature, Thomson wonders whether giving a developmental account of a feature is enough to eliminate a selective one, he seems to be expressing misgivings about this alternative causes assumption. Trying to separate causes by partitioning variance into selective and constraining components, furthermore, presumably involves at least some of the difficulties that attend ANOVA in behavioural studies.[31]

Sober has analysed attempts to apportion causal responsibility to the genes and the environment. He concludes that questions about relative causal contributions to the phenotype are not locally answerable, because an organism cannot be affected by one and not the other. One can, however, ask how much difference one or the other made to a particular set of outcomes.[32]

There is a distinct possibility that the same is true of attempts to treat constraints and natural selection as competing forces. In another article, Sober discusses disputes over the relative strength of natural selection and correlated characters. When he asserts that the disagreement is not really about the power of natural selection (which is the way it is often cast), but rather about the "power of mutation", he seems to be identifying the same confusion between local and nonlocal analyses as exists in nature-nurture arguments. What appear to be questions about relative causal contributions to a particular outcome are only intelligible if recast as questions about a class of outcomes; he suggests that one could ask "how often a trait is maintained by pleiotropy even though it is selected against, or is eliminated by pleiotropy even though there was selection for it. This is a far cry from looking at a single trait whose presence in a population is the joint product of selection and pleiotropy and asking which contributed more to its evolution" (emphasis in original).[33]

Just as developmentalists have sometimes shunned partitioning in favour of the range of possibilities compromise, so some constraints theorists have spoken of developmental laws determining the range of forms attainable by evolutionary change. Gene changes simply select from this array.[34] Treating constraints and selection as competing causes implies that they can be separated. As Charles Dyke and David Depew point out, invoking constraints requires some baseline of pure selection (see Note 20 on constraints as noncanonical causes). But "if the novelty generating process and the selection process are coupled and interactive" rather than independent, no baseline of "pure selection" can be described.[35] By the same token, it would be difficult to describe an array of possible forms prior to, and independent of, selection.

INTERACTING DUALISMS, INTERACTIVE SYSTEMS

Earlier in this essay I pointed out that the nature-nurture and the constraints-selection dichotomies are not only structurally similar, but that they are related in substance as well. They are, in fact, mutually reinforcing. Evolutionary theory maintains developmental dualism by the following logic: Evolution is typically defined by change in gene frequencies. Organisms must therefore be explicable in terms of genetic "information". Genetic transmission is thus the needle's eye through which innate characters must pass. In J.T. Bonner's arresting image, the life cycle must (usually) narrow to a single cell,[36] so inherited nature must be passed on in the zygotic DNA. Acquired characters, formed by environmental action, are excluded and thus rendered irrelevant to evolutionary change. Evolutionary theory thus seems to require developmental dualism.[37] Developmental theory, in turn, insofar as it embraces the transformational model of ontogeny (as predetermined, uniform, internally generated), reinforces dualism in evolutionary studies by legitimising this vision of autonomous change.

Resolution to the nature-nurture dichotomy involves, ironically, taking development seriously. In a discussion of dichotomies in biology, Gray[38] suggests that this resolution has been achieved in studies of behavioural development, but I think his statement should be taken less as an accurate report from the front than as a rhetorical device to prod evolutionists into action. (What biologist, after all, would be indifferent to having psychologists held up as models of intellectual sophistication?) Having spent considerable time puzzling over the persistence of dichotomous thinking among developmentalists, I am less sanguine, and am convinced that one of the many factors conspiring to maintain, and repeatedly reconstruct, the nature-nurture complex is psychologists' increasing attention to evolutionary biology.

By the reasoning outlined above, evolutionary theory not only permits, but virtually obliges the making of distinctions between inherited and acquired traits,[39] and thus the embracing of the dualistic doctrine that there are two sources of organismic forms, the genes and the environment. That reasoning is based on nondevelopmental notions of transmission. (Either transmission simply ignores development or it assumes the transformational model, and is thus nondevelopmental to the degree that it is preformationist.)

The metaphors of transmission and programmed development, then, link ontogeny to phylogeny: Innate characters, fashioned by natural selection, are passed on in the genes and reappear in each generation. "Transmission", in fact, is a metaphor for this reliable reappearance in each generation.[40] What the image of biological faxing elides is the multitude of interactive changes over time that constitute epigenetic emergence and stability. The technical language of transmission genetics is correctly employed for particular distributions of developmental products (phenotypes) in populations. *But reliable developmental courses are the prerequisites for such population patterns.* Those life courses are *assumed by* the transmission metaphor, not *explained by* it.

The alternative I have offered for these nondevelopmental notions is the metaphor of construction. Emphasis on constructive *interaction* defuses the necessity of attributing power to external artificers or autonomous internal forces. A nondichotomous view of ontogeny may prepare the way for an evolutionary theory that is synthetic in a rather different sense from the usual one. It may allow, that is, the fusion of the transformational and variational models.[41]

SYNTHESIS

If anything is transmitted (made available to the next generation), it is not phenotypic features, but developmental resources or means: genes plus the rest of a widely extended developmental system.[42] "Natures", then, are simply developing phenotypes, whether common or rare, and they emerge and change by the constant "nurture" of developmental interactions. This makes nature and nurture not internal and external causes or alternative sources of organic form, but rather developmental *products* (natures) and the developmental *processes* (nurture) by which they come into being.

Since ontogeny is not an autonomous transformational process, perhaps it is not maximally useful to think of developmental constraints as insulated, internal and necessary. And since natural selection is not an active agent choosing (or worse, shaping) passive organisms, but is instead the result of the "interpenetration of organism and environment",[43] perhaps natural selection is not best conceptualised as an arbitrary external force. Since neither model is adequate to explain change, furthermore, the partitioning compromise discussed above is again called into question. If there are problems with the very concepts of internally and externally directed change, that is, combining them in varying proportions does not resolve the problems.

Instead of opposing transformational to variational accounts, or apportioning causal responsibility between them, or giving one the role of determining the range of possibilities while the other selects from that range, we can return to the models themselves (Fig. 3) and dispose of the problematic aspect of each. Change in a collection can certainly result from change in its constituent entities (developmental systems), but the change need not be uniform or predetermined. Variant entities (developmental systems) can propagate themselves with differing frequencies, but the variants themselves need not be static. (It should be noted that developmental systems are not cleanly bounded entities, and never reproduce themselves precisely.) A more ample view of development leads to a more ample definition of evolution: Evolution is change in the constitution and distribution of developmental systems.[44] Developmental systems must be understood, not as internal to the organism, and certainly not as some cover term for genetic programmes, but rather as organism-environment complexes that change over both ontogenetic and phylogenetic time.[45]

My ambivalence about the internal constraints literature is similar to the discomfort I feel about the biological trend in psychology: I am wary of a pendulum swing along a dimension I consider inappropriate. The nature-nurture controversy has involved great confusion and led to enormous numbers of unjustified conclusions, meaningless or ambiguous statements and erroneous claims. Sometimes the problems have stemmed from a failure to distinguish the many distinct senses of terms like "biological" and "innate" and sometimes, from confusion about the process of development itself. (The examples of such empirical and conceptual problems found i Notes 1, 23, 27, 32, and 38 demonstrate the importance of conceptual issues even for those who profess scant interest in them. Given the association of internality with relative fixity, many of these problems have centred on questions about the difficulties or risks of trying to alter certain aspects of human life.)

If similar conceptual problems exist in the constraints selection literature, similar inferential problems may be present as well. The scholar who is alert to this possibility might be cautious about several things: Is quantitative language being used about a particular taxon when it is only appropriate for patterns in an array of taxa? Are local analyses, of character variation across taxa, for instance, being inappropriately used to indicate general functions? If terms like "constraints" are being used in several senses, are the senses made explicit and consistently distinguished? (Even the review in Note 13 detailed a number of meanings.) If the various senses are conflated, does this lead to evidence for one kind of constraint improperly being used to infer another, in what I have called "cross inference"? Are factors that are interdependent and changing treated as independent and static? Are certain kinds of population-niche interactions being slighted as a result? Some biologists are now examining the methodological strategies of traditional approaches and exploring alternative ones.[46]

Evolutionary fixity does not carry the political and moral implications that frequently accompany questions of developmental fixity, so erroneous conclusions about developmental constraints would seem to promise less sociopolitical mischief than do ill-considered pronouncements on human nature (or on nature in general). But given the interrelation of the two dichotomies, noted earlier, perhaps we should not be too complacent. Much of the controversy over the proper scope of natural selectionist explanation, after all, has been fuelled by dismay over sociobiological claims about human psychology and culture. The assertion that some behaviour or institution has an evolutionary explanation rather than a social one feeds right into nature-nurture assumptions.[47] The analysis presented here suggests that greater circumspection about the evolutionary arguments may be in order.

If development is to re-enter evolutionary theory, from which it has long been excluded, it should be a development that integrates genes into organisms and organisms into the many levels of the environment that enter into their ontogenetic construction. Explicit inclusion of these developmentally relevant factors, so often excluded from conventional accounts, makes clear the context dependence of questions about developmental timing, universality, fixity, spontaneity, and other characteristics that are attached to notions of autonomous, genetic control, and thus thought to be entailed by a "biological" argument. Detaching such issues from the idea of genetic programming also makes clear their heterogeneity: There are many kinds of biological argument, and they should not be discussed in a manner that lumps then together willy nilly. Nature-nurture oppositions (innate vs. acquired, biological vs. cultural, etc.) encourage the conflation of diverse kinds of evidence and conclusions. We can only investigate questions about evolutionary history, current reproductive or survival advantage, development and the causal processes by which behaviour occurs (and explore the relationships among these questions) if we first distinguish them from each other.[48]

The view of change presented here is systemic and interactive. The undirectional causation of the transformational model can sometimes be approximated by circumscribed analyses in which a single factor is isolated by controlling everything else, but the results of those analyses must eventually be reintegrated into a larger framework. When Bonner spoke of the life cycle narrowing to a single cell, he was making an important point about vital continuity, but even the zygote requires, and reliably has, the rest of a developmental system: all the developmentally relevant aspects of its world. The rest of the system shows continuity just as surely as the genes do, and as is the case for genes, continuity is often achieved through reconstructed structure and function, not through material identity.[49]

Study of the life cycle includes the investigation of these many influences and of the ways they become available at the time they are needed, whether by being supplied by conspecifics, by being regular features of the niche, by being sought or constructed by the organism itself, or by influencing development in some other way. The nucleus, cytoplasm, membrane, and the rest of the cell can pass through the eye of the reproductive needle, but the other interactants need not; what is important is that they be reliably present in the next life cycle.

Resolution of the nature-nurture dichotomy is, I am convinced, good in itself. In this sense, I agree with those theorists whose attempts at compromise and resolution I have criticised: Something was amiss. In addition, however, this resolution may also make us hesitate before pitting internal constraints against external selection when we conceptualise the emergence and persistence of form in evolution. That is, it may reduce the danger of recapitulating the sterile debate over inner and outer causes (and the multiple ambiguities, crossed inferences, and wild extrapolations that go with it) in our thinking about evolutionary stability and change.

NOTES

[1] Oyama (1981) 'What Does the Phenocopy Copy?' *Psychological Re-*

ports **XLVIII**, pp. 571-581; (1982) 'A Reformulation of the Idea of Maturation', *Perspectives in Ethology* **5**, P.P.G. Bateson & P.H. Klopfer (eds.), Plenum; (1985) *The Ontogeny of Information: Developmental Systems and Evolution*, Cambridge University Press; (1989) 'Innate Selfishness, Innate Sociality', *Behavioural and Brain Sciences* **XII**, pp. 717-718; (1989) 'Ontogeny and the Central Dogma: Do We Need the Concept of Genetic Programming in Order to Have an Evolutionary Perspective?', 'Systems and Development'. *The Minnesota Symposia on Child Psychology*,. **22**, M.R. Gunnar and E. Thelen (eds.), Erlbaum Associates.

[2] What is especially interesting here is that culture is treated as an external factor in the first case and a conservative, internal one in the second. On culture vs. geography see R.M. Netting (1978) *Cultural Ecology*, Cummings; see also op. cit. (Note 1, Ontogeny of Information) p. 174.

[3] R.D. Gray (1989) 'Oppositions in Panbiogeography: Can the Conflicts between Selection, Constraint, Ecology, and History be Resolved?', *New Zealand Journal of Zoology* **XVI**, pp.787-806. For discussions of origin of life and other biological oppositions, see papers in MW. Ho and W.W. Fox (eds.) (1988) *Evolutionary Processes and Metaphors*, Wiley. The embryological example is distinctive in that developmental regulation, while traditionally implying responsiveness of developing tissue to its immediate environment, involves the restoration of typical outcomes, not the production of variant ones.

[4] Or, more globally, "the environment". To realise just how global most references to the environment are, consider the fact that "the environment" usually means "everything in the universe except the genes", and that in carving creation into these two segments one must conceptually excise the DNA from the multitude of cells in which it resides. See S. Oyama (1990), 'Commentary. The Idea of Innateness: Effects on Language and Communication Research', *Developmental Psychobiology* **XXIII**, pp.271-277.

[5] Three of these are sketched out in Oyama (in press) 'Is Phylogeny

Recapitulating Ontogeny?', *Understanding Origin: Contemporary Ideas on the Genesis of Life, Mind and Society,* F. Varela and J.-P Dupuy (eds.), Kluwer Academic Publishers. See also the much more extensive treatments in R.D. Gray (1987) 'Beyond Labels and Binary Oppositions: What Can be Learnt from the Nature/Nurture Dispute?', *Rivista di Biologia/Biology Forum* **LXXX**, pp. 192-196; (1988) 'Metaphors and Methods: Behavioural Ecology, Panbiogeography and the Evolving Synthesis', *Evolutionary Processes and Metaphors,* MW. Ho and W.W. Fox (eds.), Wiley; and op. cit (Note 3).

[6] R.C. Lewontin (1982) 'Organism and Environment', *Learning, Development, and Culture* , H.C. Plotkin (ed.), Wiley: (1983) 'Darwin's Revolution', *New York Review of Books* , **XXX, 10, June 16,** pp. 2127; and Elliott Sober (1984) *The Nature of Selection* , Bradford/MIT Press; (1985) 'Darwin on Natural Selection: A Philosophical Perspective', *The Darwinian Heritage* , D. Kohn (ed.), Princeton University Press. See also R. Levins and R. Lewontin (1985) *The Dialectical Biologist* , Harvard University Press; and S. Oyama (1988) 'Stasis, Development and Heredity', *Evolutionary Process and Metaphors* , MW. Ho and W.W. Fox (eds.), Wiley.

[7] See op. cit. (Note 1, 'Maturation'). Though the transformational model is formulated to explain change in a collection of entities, the focus here is, of course, on change in an individual organism. One could think of organismic development in terms of transformational change at the level of organs or tissues, but this would involve attributing the higher-level development to a heterogeneous assemblage of subsystems, and tissues exert a great deal of influence on each other. The organisation among the subsystems would then present additional problems for a straightforward transformational account.

[8] By this I mean only statistical typicality; no essential species nature is required for such probabilistic generalisations to be made. See discussion of programmed development in op. cit. (Note 1, Ontogeny of Information and 'Central Dogma').

[9] J. Morss, this volume; (1990) *The Biologising of Childhood: Develop-*

mental Psychology and the Darwinian Myth, Erlbaum Associates. In op. cit. (Note 6, 'Darwin'), in fact, Sober refers to the model of internally driven change as "developmental" rather than "transformational".

[10] See B.F. Skinner (1981) 'Selection by Consequence', *Science*, **CCXIII**, pp.501-504; on neural selection see G.M. Edelman and V.B. Mountcastle (1978) *The Mindful Brain*, Bradford Books/MIT Press; on the immune system see N.K. Jerne (1967) 'Antibodies and Learning: Selection versus Instruction', *The Neurosciences: A Study Programme*, **1**, G. Quarton, T. Melnechuk and F.O. Schmitt (eds.), Rockefeller University Press; on language, see K. Wexler and P. Culicover (1980) *Formal Principles of Language Acquisiton*, MIT Press; and for a sweeping view of selectionist explanation, as well as many references, see Massimo Piattelli-Palmarini (1989) 'Evolution, Selection and Cognition: From "Learning" to Parameter Setting in Biology and in the Study of Language', *Cognition*, **XXXI**, pp. 144. It should be noted that there is a difference between selection from an actual array of objects or responses (organisms, neurons, operants) and selection as parameter setting, a difference that tends to be ignored when selectional models from neurobiology or immunology are mustered as support for the nativist project in the cognitive sciences.

[11] This has the qualities of a good origin myth; the story is more complex. For more appreciative views of Lamarck, see MW. Ho and P.T. Saunders (1982) 'Adaptation and Natural Selection: Mechanism and Teleology' *Towards a Liberatory Biology*, S. Rose (ed.), Allison and Busby; and P.J. Taylor (1987) 'Historical versus Selectionist Explanations in Evolutionary Biology', *Cladistics*, **III**, pp. 113. For more historical detail, see H.E. Gruber (1981) *Darwin on Man: A Psychological Study of Scientific Creativity*, 2nd ed., University of Chicago Press; and L.J. Jordanova (1984) *Lamarck*, Oxford University Press.

[12] But see J.R. Grehan and R. Ainsworth (1985) 'Orthogenesis and Evolution', *Systematic Zoology*, **XXXIV**, pp.174 - 192.

[13] J. Maynard Smith, R. Burian, S. Kauffman, P. Alberch, J. Campbell, B. Goodwin, R. Lande, D. Raup and L. Wolpert (1985) 'Developmental

Constraints and Evolution', *Quarterly Review of Biology*, **LX**, pp. 265 - 287.

[14] Behaviourists have explicitly compared operant shaping with biological evolution; See op. cit. (Note 10, 'Selection by Consequence'); R.J. Herrnstein (1989) 'Darwinism and Behaviourism: Parallels and Intersections', *Evolution and its Influence*, A. Grafen (ed.), Clarendon Press.

[15] S.J. Shettleworth (1972) 'Constraints on Learning', *Advances in the Study of Behaviour*, **IV**, pp. 168, Academic Press; R.A. Hinde and J. Stevenson-Hinde (eds.) (1973) *Constraints on Learning*, Academic Press; on genes and society, see E.O. Wilson (1978) *On Human Nature*, Harvard University Press.

[16] Or "ahistorical universals", S.A. Kauffman (1985) 'Self Organisation, Selective Adaptation, and Its Limits: A New Pattern of Inference in Evolution and Development', *Evolution at a Crossroads: The New Biology and the New Philosophy of Science*, D.J. Depew and B.H. Weber (eds.), Bradford Books/MIT Press, p.171. Similarly, Ho and Saunders declare that a "a scientific study should consist in the delimitation of the necessities which underlie the process of evolution, without recourse to contingencies" (p.590, emphasis in original), M.W. Ho and P.T. Saunders (1979), 'Beyond neoDarwinism An Epigenetic Approach to Evolution', *Journal of Theoretical Biology*, **LXXVIII**, pp. 573 - 591. See also B.C. Goodwin (1982) 'Biology without Darwinian Spectacles', *Biologist*, **XXIX**, pp. 108 - 112.

[17] See op. cit. (Note 13) on universal versus local constraints.

[18] See op. cit. (Note 6, 'Stasis').

[19] P. Alberch (1980) 'Ontogenesis and Morphological Diversification', *American Zoologist*, **XX**, pp. 653 - 667; S.J. Gould (1989) 'A Developmental Constraint in Cerion, with Comments on the Definition and Interpretation of Constraint in Evolution', *Evolution*, **XLIII**, pp. 516 - 539; see op. cit. (Note 13). The subtitle of the Hinde and Steven-

sonHinde volume on constraints on learning is "Limitations and Predispositions", op. cit. (Note 15).

[20] S.C. Stearns (1986) 'Natural Selection and Fitness, Adaptation and Constraint', Pattern and Process in the History of Life, D.M. Raup and D. Jablonski (eds.), Springer-Verlag; ibid., 'Developmental Constraint in Cerion', p. 516.

[21] See op. cit. (Note 12). See also op. cit. (Note 16, 'Biology without Spectacles') p. 112; similarly, Ho and Saunders "place more emphasis on the physiological and developmental potential (or internal factors) of the organisms as opposed to the 'external' factors of random mutation and natural selection", op. cit. (Note 16) pp. 589 - 590.

[22] The relation between artist and materials is being re-thought, even in the popular press. Of a recent exhibit of Japanese sculpture a news-magazine reports: "the exhibition rightly contends that its artists (or any artists, if you think about it) don't transform their materials so much as redirect them. They don't make everlasting objects out of inert and characterless stuff. . . Instead, they highlight a few inherent qualities of their materials for a relatively brief moment in time." P. Plagens (1990) 'Palms and Circumstance', *Newsweek*, **August 20**, p. 64.

[23] S. Oyama (in press) 'Bodies and Minds', *Journal of Social Issues, special issue on evolution*, M. Brewer and L. Caporael (guest eds.).

[24] K.M. Noonan (1987) 'Evolution: A Primer for Psychologists'; and R. Thornhill and N. M. Thornhill (1987) 'Human Rape: The Strengths of the Evolutionary Perspective', both in *Sociobiology and Psychology*, C. Crawford, M. Smith and D. Krebs (eds.), Erlbaum Associates.

[25] See op. cit. (Note 1, Ontogeny of Information and 'Central Dogma').

[26] For a classic critiques of nature-nurture dichotomising, see D.S. Lehrman (1953) 'A Critique of Konrad Lorenz's Theory of Instinctive Behaviour', *Quarterly Review of Biology*, XXVIII, pp. 237 - 363; (1970) 'Semantic and Conceptual Issues in the Nature-Nurture Problem',

Development and Evolution of Behaviour, L.R. Aronson, E. Tobach, D.S. Lehrman, and J.S. Rosenblatt (eds.), Freeman. The distinction between genetic and environmental information is associated with K. Lorenz (1965) *Evolution and Modification of Behaviour*, University of Chicago Press. For critiques of this partitioning of information see op. cit. (Note 1, Ontogeny of Information) and T.D. Johnston (1987) 'The Persistence of Dichotomies in the Study of Behavioural Development', *Developmental Review*, **VII**, pp. 149 - 182.

[27] R. Lewontin (1974) 'The Analysis of Variance and the Analysis of Causes', *American Journal of Human Genetics*, **XXVI**, pp. 400 - 411.

[28] See op. cit. (Note 1, 'Central Dogma'). See also op. cit. (Note 3, 'Oppositions') p. 803, where Gray says that constraints are considered primary; though he and I read the literature differently, neither of us likes the insistence on designating one cause as dominant.

[29] See E.F. Keller (1985) *Reflections on Gender and Science*, Yale University Press, on the language of domination in science; and J. Sapp (1987) *Beyond the Gene: Cytoplasmic Inheritance and the Struggle for Authority in Genetics*, Oxford University Press, on the battles over the relative importance of the nucleus and the cytoplasm in development. The combatants in the nucleus-cytoplasm conflict deployed some of the same rhetorical strategies I describe here; there are insides and outsides even within the cell membrane. Sapp offers interesting comments on the relationship between geneticists' descriptions of nuclear processes and their conceptions of themselves.

Though my comments here can be judged within the orthodox frameworks of developmental and evolutionary theories, they arise from a somewhat different one. It is possible to argue for an alternative framework without lapsing into rank relativism. To do so, however, involves some rather serious thinking about science: See Helen Longino's (1990) *Science as Social Knowledge*, Princeton University Press.

[30] P. Alberch (1982) 'The Generative and Regulatory Roles of Development in Evolution', *Environmental Adaptation and Evolution*, D. Mossakowski and G. Roth (eds.), Gustav Fischer characterises the developmental generation of new bauplane as proceeding "autonomously

from external environmental factors" (p. 23), and declares that "the evolution of developmental systems is characterised more by the internal structure of the developmental programme than by the external evolution of the environment" (p.25). (See Note 45.) See also op. cit. (Note 10); Stearns asks about the relative importance of internal and external factors.

[31] K. Thomson (1985) 'Essay Review: The Relationship Between Development and Evolution', *Oxford Surveys in Evolutionary Biology*, **II**, pp. 220 - 233. See Stearns op. cit. (Note 20) on attributing variance to constraints and to selection; for discussion and references see op. cit. (Note 3, 'Oppositions').

[32] E. Sober (1988), 'Apportioning Causal Responsibility', *Journal of Philosophy*, **LXXXV**, pp. 303 - 318. See also op. cit. (Note 26).

[33] E. Sober (1987) 'What is Adaptationism?', *The Latest on the Best*, J. Dupre (ed.) MIT Press (p. 115). Interestingly enough, Stearns asserts that selection and constraint are involved in all evolution, but that the problem is to determine their relative influence, see op. cit. (Note 20). Behavioural scientists justify their continued pursuit of genetic and environmental "components" in precisely the same way.

Sober refers to the "power of mutation". "Mutation" can refer both to alterations in DNA sequences and to the phenotypic consequences of such alterations. Restrictions on the range of phenotypic results of genetic mutations are surely a matter, not just of constraints on DNA changes, but of the rest of the developmental systems in which they occur. A particular DNA change may have no effect in some systems, and a variety of effects in others. The outcome will depend both on the alteration and on the rest of the system. It might thus be more apposite to speak, as I have in this essay, not of the power of natural selection or of mutation, but of the possibilities for variation in developmental systems. Whether any particular phenotypic variation will occur is a function of the system dynamics, and it is the dependence of such variation on this interactive complex that is indexed (but not captured) by the notion of mutational power.

It might be wondered (and has been wondered, by Kim Sterelny, in personal communication, 1990) whether this is just another version of the imperialistic move. The short answer is no. A slightly longer one is: Sort of, but not really: Genetic imperialism seeks to decontextualise gene action by collapsing all possible ontogenetic outcomes into some notion of "genetic information", while the developmental systems formulation makes contextual dependence explicit by stressing joint determination of outcome by the system and by its perturbation. The real imperialistic move for a developmental systems theorist would be to claim that a system "determines" all of its possible changes prior to specification of the particular perturbation. (Something like this move is documented in the next note; notice, however, that development is autonomous and internal in those theories,not interactive).

Any outcome of a multiplicative function is specified by one factor, given the other one(s). Genetic imperialism gives the genes the power to specify all outcomes given only themselves; this is like saying that 2 "specifies" the products of all multiplications in which it might possibly be a multiplier, and that it does so before the fact; the multiplicand simply selects from this prior array. This is a most peculiar claim; read for its rhetorical function, it is seen to be a ploy to make certain causes recede into the background. Provided they are understood amply enough, developmental systems block this move for either "internal" or "external" factors by including them both. If arbitrary causal domination is abolished within a system's boundaries, is it still an empire?

[34] G. Webster and B.C. Goodwin (1982) 'The Origin of Species: a Structuralist Approach', *Journal of Social and Biological Structures*, V, pp. 15 - 47; op. cit. (Note 16, 'Biology without Spectacles'). See also op. cit. (Note 19, 'Ontogenesis') p. 664. Alberch says that development is "crucial" in that "it defines the realm of the possible". Significantly, in describing macroevolution as an interaction between "production of morphological novelties (epigenetically determined) and differential extinction (environmentally determined)" (p. 664), he maintains the classic internal-external dichotomy, in which epigenesis is an internal process, independent of the environment, and selection, an external one, independent of development.

[35] C. Dyke and D. Depew (1988) 'Should Natural Selection be an Explanation of Last Resort? Well, Maybe not the Last Resort, but ....', Rivista di Biologia , *Biology Forum* , **LXXXI**, pp. 115 - 129 (p. 117).

[36] J. T. Bonner (1974) *On Development*, Harvard University Press. Bonner emphasised the importance of other constituents of the germ cell, maintaining that focus on nuclear DNA was too narrow. I agree, but find no warrant for stopping at the cell wall.

[37] S. Oyama (in press) 'Transmission and Construction: Levels and the Problem of Heredity', *Critical Analyses of Evolutionary Theories of Social Behaviour: Genetics and Levels.* Monograph I of the T.C. Schneirla Conference Series, G. Greenberg and E. Tobach (eds.), Shapolsky Publishers; see also op. cit. (Note 1, 'Central Dogma').

[38] See op. cit. (Note 3, 'Oppositions'). See also P. Bateson (1983) 'Genes, Environment and the Development of Behaviour', *Animal Behaviour* , **3**, *Genes, Development, and Learning,* T.R. Halliday and P.J.B. Slater (eds.), Blackwell; see also op. cit. (Note 26).

[39] Though the distinction can be made without any specific assumptions about the dynamics of developmental processes, it seldom is; in fact, it is typically treated as a statement about developmental mechanism.

[40] See op. cit. (Note 37).

[41] See op. cit. (Note 5, 'Metaphors' and 'Phylogeny Recapitulating Ontogeny': Note 6, Dialectical Biologist and 'Stasis').

[42] See J. Cohen (1979) 'Maternal Constraints on Development', *Maternal Effects in Development* , D.R. Newth and M. Balls (eds.), Cambridge; see also op. cit. (Note 1, 'Maturation', Ontogeny of Information).

[43] R.C. Lewontin, S. Rose and L.J. Kamin (1984) *Not in Our Genes* , Pantheon, speak of "codevelopment of the organism and its environ-

ment" (p. 275); see also op. cit. (Note 6, Dialectical Biologist).

[44] Oyama op. cit. (Note 6).

[45] They must therefore be distinguished from other uses of the phrase, such as Alberch's decidedly internalist "developmental programmes" op. cit. (Note 19; see also Note 30) and his "developmental systems" (1982) 'Developmental Constraints in Evolutionary Processes', *Evolution and Development* , J.T. Bonner (ed.), SpringerVerlag, as well as from other less extended conceptions of ontogeny. Similarly, the definition of heredity presented here is significantly broader than the "hereditary apparatus" of MaeWan Ho (1984) 'Environment and Heredity in Development and Evolution', *Beyond neo Darwinism: An Introduction to the New Evolutionary Paradigm* , MW. Ho and P.T. Saunders (eds.), Academic Press. Consisting of the nucleus and cytoplasm, this "apparatus" is considerably more restricted than the formulations in her later papers, where she appears to argue for the inheritance of something like my developmental system: for example (1988) 'On Not Holding Nature Still', *Evolutionary Processes and Metaphors* , MW. Ho and S.W. Fox (eds.), Wiley.

T.D. Johnston and M.T. Turvey (1980) 'A Sketch of an Ecological Metatheory for Theories of Learning', *The Psychology of Learning and Motivation* , **14**, G.H. Bower (ed.), Academic Press, capture the same interactive complex when they speak of the "co-implicative" relationship between organisms and their surrounds (p. 152), as does R.D. Gray with his "reciprocally constrained construction", in (1987) 'Faith and Foraging: A Critique of the Paradigm Argument from Design', *Foraging Behaviour* , A.C. Kamil, J.R. Krebs and H.R. Pulliam (eds.), Plenum Press. B.C. Patten's "environs" may also refer to the same sort of inclusive complex: (1982) 'Environs: Relativistic Elementary Particles for Ecology', *American Naturalist* , **CXIX**, pp. 179 - 219. Whether these authors focus on development or on function (as in this last group of works), what they have in common is an interest in reducing the conceptual distance between organisms and their surroundings. See op. cit. (Note 11, 'Historical versus Selectionist Explanations'; Note 43).

[46] P.D. Dwyer (1984) 'Functionalism and Structuralism: Two Programmes for Evolutionary Biologists', *American Naturalist*, **CXXIV**, pp. 745 - 750; Gray ibid. and op. cit. (Note 3, Note 5 'Beyond Labels' and 'Metaphors'; ibid. 'Environs'; op. cit. (Note 11, 'Historical versus Selectionist Explanations'). See also Note 43.

[47] See op. cit. (Note 1, 'Innate Selfishness'; and Note 23). Herrnstein asserts, "Nature versus nurture in regard to behaviour is the last great evolutionary controversy" op. cit. (Note 14) p. 40.

[48] N. Tinbergen (1963) 'On aims and methods in ethology', *Zeitschrift fuer Tierpsychologie*, **XX**, pp. 410 - 433.

[49] The continuity of the germ line, that is, is achieved by repeated reconstruction of DNA strands, and the continuity of other developmental interactants may also involve reconstruction rather than simple persistence.

Susan Oyama,
Department of Psychology,
City Univestity of New York.

John R. Morss

AGAINST ONTOGENY

The notion of a predictable sequence of developmental states, general across individuals of a species, is central to much of developmental biology. Accounts of such predictable developmental change in an individual ('ontogeny') have played an important role in evolutionary theory over many years. Charles Darwin, for example, appealed to embryological evidence to support his claims concerning descent with modification. More recently, notions of ontogeny have played a central part in the 'heterochrony' formulation of Stephen Jay Gould. In this paper I present three arguments against current uses of the notion of ontogeny. The first argument, grounded in the study of human development, suggests that current versions of ontogeny derive from non-Darwinian evolutionist biology. The second argument extends the analysis of Alberch[1] in his critique of the stage-sequence version of ontogeny within the contemporary life sciences. The third argument derives from the recent arguments of Hull[2] concerning the biological concept of the individual, and suggests that current versions of ontogeny depend on a natural-kind formulation of the individual organism which is not valid.

The notion of ontogeny is intimately linked with three powerful terms, 'evolution', 'development' and 'progress'[3]. As Gould has observed, Darwin avoided the term evolution because of its connotations of progressive development, connotations embraced with great enthusiasm by Herbert Spencer. The restriction of the term development to individual lifespans is a recent convention. My argument in this paper is that ontogeny, taken as a feature of what we now call development, is a characterisation of change which remains committed to the evolution-development-progress complex. Like the more classic, Haecke-

P. Griffiths (ed.), Trees of Life, 241–269.

lian versions of 'phylogeny', available versions of 'ontogeny' connote predictable, meaningful, progressive change through an ascending series of states. If my analysis is justified, then only a casual, informal sense of ontogeny could remain - just as only a casual, informal sense of phylogeny is now in currency. The resounding Haeckelian phrase 'ontogeny and phylogeny' would become limited to the titles of books and of conference papers.

For a combination of reasons, biologists have left the notion of ontogeny largely unexamined. Developmental biologists have studied fine details and mechanisms more than the conceptual framework itself[4]. Ontogeny has been more closely scrutinised within developmental psychology. The analyses undertaken by those psychologists interested in human development might be of very direct relevance to quite important issues in evolutionary biology at large. As indicated in the first section of this chapter, the investigations of developmental psychologists into their own subject matter have cast doubt on the utility of the notion of ontogeny in the human case. The historical investigations of these same scholars have shown that developmental psychology represents the application or working-out of a research programme grounded in non-Darwinian evolutionist biology[5]. That is to say, the biology on which developmental psychology rests is itself committed to the progressive evolutionism connoted by Haeckel's twin terms phylogeny and ontogeny. Developmental psychology is Haeckelian, and Spencerian, much more than it is Darwinian.

These conceptual investigations by developmental psychologists therefore cast doubt both on the validity and the utility of their notion of ontogeny. The notion of ontogeny would also seem to be supported by the contemporary model of the genetic programme. However, serious problems have been identified with the notion of a genetic programme as an explanatory entity implicated in the usual conception of a fixed developmental sequence[6]. To a considerable extent, the genetic programme has itself been validated by an appeal to ontogeny; that is, by appeal to an observed sequence in development which would appear to require a genetic base. Problems with either genetic programmes or ontogeny mean problems for the other; convergent problems with both give rise to very serious consequences.

Following the first section of this paper, which focuses on recent scholarship within psychology, the second section discusses the ways in which the life sciences currently employ and understand ontogeny. In general, ontogeny is understood in terms of a fixed series of states or stages. Even such unorthodox voices as that of Gould retain this traditional version of ontogeny. However, this position has already been criticised from within developmental biology, most notably by Alberch.

The implications of the analysis of ontogeny are serious for both biologists and psychologists of development. There are some closely related issues of concern to both disciplines. Their common focus of interest would appear to be change with identity - the processes of change undergone by some underlying substrate over a period of time. In a naive sense at least, developmental biology and developmental psychology would appear to be historical sciences - that is, studies of patterns (or at least sequences) of events over time. History as a discipline has itself, at times, endorsed a transcendental orientation in which events are seen as unfolding in a linear and progressive manner, much like the ontogeny model itself. Of course, organic models for history can be traced back to common origins with stage models for living things - in the writings of Herder, for example[7]. But contemporary approaches to history seek to avoid such progressive interpretations (one kind of which is often termed 'Whiggism').

The contemporary debate on the role of the 'historical entity' in biology, especially following the writings of David Hull and Michael Ghiselin, is therefore of considerable interest. This is the topic of the third section. If the notion of a species as an individual and the related concept of historical entity are legitimate ones, then an alternative framework to ontogeny might be available. History, in this sense, might prove to be a more appropriate discipline for the investigation of those phenomena we currently subsume under the rubric of 'development'.

The critical examination of ontogeny shows that it is insufficient for its present explanatory role in biology and psychology. In the absence of this ontogenetic version of development we might seek for an historical one. Ontogeny is a ladder, and a ladder represents an inappropriate and alien appendage to a tree of life.

1. ONTOGENY IN DEVELOPMENTAL PSYCHOLOGY: A NON-DARWINIAN VESTIGE

The science of developmental psychology was established towards the end of the last century[8]. It began at a time of intense interest in evolution, broadly defined, of which Darwinism was but one version[9]. The writings of Haeckel and of Spencer were of equal, perhaps greater impact on early developmentalists than were the writings of Darwin. It was Haeckel who first introduced the terms ontogeny and phylogeny, defining them in terms of their interrelationship in the evolutionary process[10]. It was Spencer who first applied a thoroughly evolutionary approach to developmental phenomena in general, including what we now call 'individual development'. Developmental psychologists took their idea of progressive, recapitulationist developmental stages from the biology of their time. To the extent that progressive recapitulationism has been abandoned in biology, developmental psychologists can claim no biological support for their ideas.

In many ways developmental psychology represents the application or working-out of a research programme grounded in this peri-Darwinian milieu. More precisely, the research programme has been non-Darwinian - despite the discipline's self-image. The versions of evolution which most strongly influenced the early developmentalists were those which stressed order, purpose, and direction in large-scale change and the origins of species. Such mechanisms as the recapitulation of ancestry in the individual, and the inheritance of acquired characters, were consistent with this 'progressive' account of evolutionary change. Both of these supposed mechanisms tied individual development closely to large scale evolution. To a considerable extent, indeed, studies of individual development could be seen as a testing ground for theories of evolution. The developing child, after all, is evolving 'before your very eyes'.

Influential research into children's development was carried out by physiologist and embryologist Wilhelm Preyer in Germany, and by psychologist James Sully in England, in the 1880s and early 1890s. Various aspects of early mental development were investigated. Both Preyer and Sully were personally acquainted with Darwin (Preyer writing a biography[11]) and both made direct appeal to evolutionary arguments. A third contemporary was George John Romanes who published accounts of child development in 1888, and had very direct links with Darwin. The writings of all three - Preyer, Sully, and Romanes - in spite of their nominal connections with Darwin, emphasised a variety of factors which are now considered non-Darwinian: 'Lamarckian'-style inheritance of the experience of ancestors and direct recapitulationary relationships between ontogeny and a supposed linear sequence of ancestral species-forms. As observed by Preyer, for example:

"The mind of the new-born child, then, does not resemble a tabula rasa ... the tablet is already written on before birth, with many illegible, nay unrecognisable and invisible, marks, the traces of the imprint of countless sensuous impressions of long-gone generations ... each [man] must ... fill out and animate anew his inherited endowments, the remains of the experiences and activities of his ancestors."

Preyer[12] states elsewhere that

"individual development ... is an abbreviated repetition of phylogenesis ... the mental development of all mankind can be found in the child in an abbreviated form."

It must be noted that Darwin himself did in fact flirt with these 'non-Darwinian' mechanisms, and indeed *The expression of the emotions* is "almost entirely"[13] based on Lamarckian-style inheritance (apologetics from contemporary psychologists notwithstanding). The strength of Darwin's position is now taken to be his alternative to such mechanisms. The matter is somewhat academic in light of Darwin's slight impact on the actual research of early developmentalists. For example, Darwin's own developmental observations (of his infant son) seem to have had very little influence on the emerging science of developmental psychology[14].

The first generation of developmental researchers appealed to Darwin for legitimacy - for the validity of looking at early development within what they took to be a modern, evolutionary context. But the substance of their explanations for developmental process and characteristics were not Darwinian ones. Much more popular than what we now discern as Darwinism was a view of evolution in terms of progress, transcendental order, and hierarchy. The adoption of this non-Darwinian mode might perhaps be put down to deficiencies in the psychological scientific community - that is, it might be argued that the psychologists simply misunderstood Darwin. But the misunderstanding was not limited to psychologists, as has been shown by the writings of Peter Bowler[15]. This 'non-Darwinian' perception by the psychologists was entirely typical of the scientific (including lay-scientific) reception of Darwinism. The ways in which psychologists began to look at the growth of the individual mind were consistent with prevailing attitudes in anthropology, for example[16].

One very important set of claims, consistent with these non-Darwinian assumptions, was for the legitimacy of comparisons between (civilised) children and various kinds of inferior human. The recapitulation framework suggested valid comparisons between children of 'civilised' races (who would still develop further) and the adults of 'primitive' races, who had reached their ceiling of personal development. On different occasions and in different ways, parallels were drawn with prehistoric man as reconstructed; with savages; with women; with the labouring class; with the Irish; with the insane; with the dreamer; as well as with various kinds of animal. This comparative programme was to prove very fruitful in the hands of Freud, as well as Piaget, for example. Its importance was (and is) much wider than the study of development per se. Its widespread influence highlights the importance of the hierarchical nature of a stage theory of development.

Various forces, in addition to evolutionary biology, were at work in the latter 19th century in the encouragement of certain kinds of child study. Education was seen by several influential traditions (including the Spencerian and the Herbartian) as needing to be modelled on the 'natural' course of development, which therefore needed to be studied - in order for the curriculum to be scientifically grounded. The natural course of development was defined in terms of a series of stages. In turn, the stages of individual growth were taken to correspond with stages in the history of western civilisation. In the USA, there was considerable public and professional pressure for the instigation of academic 'child study' - and in the 1880s this demand was recognised by Stanley Hall who fairly rapidly became a world figure in the psychology of development with particular reference to middle childhood and adolescence. Hall's evolutionism was, again, entirely non-Darwinian (in retrospect). It was based on recapitulation and Lamarckian inheritance[17]. For example, his very influential account of adolescence, first published in 1904, was based on the notion that adolescence is a special stage of development corresponding to an ancestral event in which "old moorings were broken and a new level attained". Similarly, the modern six-year-old child shows signs corresponding to the sexual maturation of a (pigmy) ancestor, such signs being described by Hall in a memorable phrase as "the ripple marks of an ancient pubic beach".

The alternative developmental voice to Hall's in the USA was that of the philosopher James Mark Baldwin, writing especially in the 1890s. Baldwin had a much more technical awareness of Darwinism - that is, of the implications of a post-Weismann, neo-Darwinism - than any preceding (or perhaps any following) developmentalist. But instead of leading him into a neo-Darwinist developmental psychology, it led him to look for an alternative process to Lamarckian-style inheritance but one which could give rise to somewhat similar consequences. This was the 'Baldwin effect' which allows personal adaptation to prepare the way forward for heritable kinds of evolutionary change (a process still cited in sociobiological literature[18]). Although he disavowed the very mechanical kind of recapitulation of Haeckel and Hall, Baldwin retained the general notion of parallels between the supposed development of human society/civilisation and the development of the modern child. Consistent with this general orientation, individual development was defined in terms of a set of stages.

Contemporary with Baldwin, Sigmund Freud's account of the development of the individual was, again, based on Haeckelian recapitulation and Lamarckian-style inheritance, as has been demonstrated by Sulloway[19]. The Oedipus story as it relates to male development and the emergence of superego is a good example. The notion of stage-like progression in development is central to Freud, and is very clearly derived from an evolutionary theory: thus, according to Freud, the anatomical location and embryological trajectory of the mouth and anus in various animals was evidence for the oral-anal sequence in ontogeny. Individual development for Freud is uniform and biologically determined in its basic sequence; personality differences emerge from perturbations in the basic sequence, not from unique experiences as such. Unique experiences, at most, can switch a person's lifecourse from one preset pathway into another. The landscape of possibilities is laid out in advance.

Probably the most influential developmental psychologist this century - certainly in educational practice - has been Jean Piaget, Swiss zoologist (1896-1980). True to French-speaking biology, Piaget resisted Darwinism and it has been argued that he never really grasped its basic tenets[20]. Piaget endorsed the same kind of loose recapitulationism as Baldwin, as well as adhering to Lamarckian claims, hence arguing for a directive role for adaptation (that is, personal accommodation in the individual) in the evolutionary process. In 1976 Piaget published *Behaviour, the motor of evolution*[21]. For Piaget, development from infancy to adulthood comprises a set sequence of stages. This sequence has a rather loose relationship with a supposed sequence of phylogeny/human prehistory, but it is important to note that Piaget does endorse an essentially Haeckelian recapitulationism.

More specifically, Piaget used a recapitulationary stage theory to guide his research programme. His work on children's perspective-taking (in which children must identify what a scene looks like from another direction) was designed as a test of such a formulation[22]. Piaget's account proposed a repeating series of stages, such that children must work through the series at a certain level but then work through the same series again at a higher level. In the case of perspective-taking, Piaget argued that a sequence of stages in the perceptual development of the infant would be repeated in the cognitive development of the child. The particular experiment, with its three mountain layout, was designed to test this hypothesis. Piaget reported (in a rather misleading fashion) that children's errors on this task were indeed equivalent to the perceptual errors of babies. This particular research finding has been taken as the best empirical support for Piaget's 'egocentrism' as the overriding character of children's thinking in the 'pre-operational' stage. This claim has only been subject to empirically-based criticism within the last twenty years.

These early and classic formulations in developmental psychology were thus based on a progressive, directional evolutionism. The account of the developmental career of an individual, or ontogeny, was modelled on the supposedly established sequence making up the evolutionary past, or phylogeny. The validity of this search for a regular developmental pattern in the human species was itself unquestioned. Indeed, developmental psychology seemed to define itself in terms of such a project. Phylogeny, in an Haeckelian sense, became the model for ontogeny. Even when the notion of a causal recapitulation became discredited, therefore, a disguised phylogeny remained in operation. It is possible that phylogeny in this classical sense was itself derived, perhaps informally, from some notion of human growth. Be that as it may, Haeckelian phylogeny was certainly recapitulated by students of developmental psychology this century. In the forms established by Freud, Piaget and many others, human ontogeny was defined in terms of the regular progression through a hierarchically-ordered set of states or stages. The image, in essence, was of a ladder, or of a certain kind of tree - that kind in which the trunk represents the main line of progress, with the most developed point at the very top.

Contemporary approaches to human development within psychology include several explicitly biological approaches. It can be shown that these approaches - such as the ethological and the psychobiological - retain commitments to the non-Darwinian, progressive evolutionism of the classic formulations. For example, Konrad Lorenz' non-Darwinian commitments and sources of influence - especially the Haeckelian Bölsche - undermine the credibility of ethology as the basis for a Darwinian developmentalism. Lorenz' view of evolution involved a hierarchical sequence of animal types, with humans definitely at the top[23]. Non-Darwinian claims - including gross adaptationism, and flirtation with recapitulationism - may be found in the writings of contemporary ethologists of development[24]. It is still commonplace for psychologists to refer to 'the phylogenetic scale' despite Hodos and Campbell's analysis of that practice[25]. Biologically based contemporary approaches, now including developmental sociobiology, retain the commitment to ladder-like hierarchies of animal types, and hence of stages within a lifespan.

Alongside the technical and methodological progress of the discipline of developmental psychology this century, the science of genetics has likewise risen to greater and greater heights. Psychologists interested in development have been able to assume that a genetic basis for developmental instructions resides in the nuclear material. Some kind of programme is therefore said to be present within the DNA, capable under certain circumstances of generating the ontogenetic career appropriate to that species. The validity and utility of this form of explanation has recently come under serious attack. The genetic material is defined in terms of its function within the predictable development of the organism. It is a point of origin for adult morphologies and behaviours, conveniently located within the nucleus of each somatic cell. Oyama argues that the status of the genetic programme rests on such arguments rather than on independent grounds of empirically demonstrated causal relationships[26]. If Oyama's critique is even partly valid, then the notion of a genetic programme loses much if not all of its explanatory power. It could certainly not be used as independent grounds for the existence of regular patterns in development, i.e. ontogeny.

Developmental psychology, over the last one hundred years, has comprised a working-out of the prevailing non-Darwinian evolutionism of the latter end of the last century. It has maintained the same basic commitments - to order and progress in developmental change. The notion of ontogeny, at least in the hands of developmental psychologists, has encapsulated these commitments[27]. Ontogeny is defined as an ordered series of stages: steps on a ladder which reaches from egg to adult. Since Darwin, it has seemed plausible for each species to possess its own ladder. Some contemporary developmental biologists argue that it is these species-specific ladders which best define the species itself. An analysis of the history of developmental psychology shows that the notion of ontogeny is little more than the application of evolutionist thinking to the individual lifespan. Developmental psychology cannot offer any support to current notions of ontogeny, except the support of a century-old tradition. If ontogeny is a valid notion, then its validity will have to be demonstrated within biology.

2. ONTOGENY IN THE LIFE SCIENCES : AN ELASTIC CONCEPT

This section presents an overview of the ways in which ontogeny is utilised in contemporary biology. Particularly detailed consideration is given to the work of Stephen Jay Gould. It is Gould's heterodox status that makes his writing important in this context. Gould has mounted a well-coordinated attack on progressivism in paleontology, and yet his own appeal to ontogeny is itself progressivist. While many of Gould's pronouncements on macroevolution may be heterodox, his position on ontogeny is remarkably traditional. Next, Alberch's criticisms of stage theories of ontogeny are considered, and some putative alternatives noted. Finally, brief consideration is given to the related notion of 'developmental pathway'.

References to ontogeny or to ontogenetic stages may be found in a variety of contemporary sources in the life sciences. Rosen[28] for example in discussing speciation in teleost fish, outlines a sequence of states for the jawbone. The sequence represents ontogenetic development, but each state can also be an adult form and can hence characterise a species. Juvenile states in one organism are taken as identical with adult states in a supposedly related organism, as in classical recapitulation. Eldredge and Cracraft[29] refer to ontogeny in a similar case of sequences of bone and cartilage in the skeleton. Again, the juvenile cartilaginous state in one organism is taken to correspond with the adult cartilaginous state of a related organism. For Rosen, ontogeny is important since it provides direct information on 'hierarchies' in nature, and by 'hierarchy' he means a progressive, ordered series. 'Ontogenetic hierarchy' is used in a somewhat different sense by Fortey and Jefferies[30], but in a sense which shares the assumption of order in a sequence of states.

In his version of cladistics, Janvier[31] emphasises the explanatory power of the sequence of character states: 'the ontogenetic criterion' is for Janvier the 'only reliable criterion' in locating true relationships. That is, a state which occurs earlier in one species than in another can provide firm evidence on common ancestry. Clearly, Janvier's procedure relies on the validity of his analysis of state-sequences, and on the validity of equating states across species. Similar arguments in the writings of Nelson and Patterson are noted by Alberch[32], who observes that "the assumption that ontogeny is ordered and that this order is retrievable is the cornerstone of their analysis".

All these usages of ontogentic sequences rely on the identity between states which are adult in one species and juvenile in another. This has been described by Alberch as an appeal to a kind of homology. Essentially, character states are treated as quasi-autonomous units which can be strung together in a sequence. According to this model, states which formerly characterised the adult come to characterise the juvenile, as a result of the addition of new terminal states. Different species can be pictured as made up of varying sections of the whole chain, usually including the common starting point. This picture, when applied to the embryo, gives rise to the classical (Haeckelian) recapitulation account of embryogenesis. It is usually argued that some Von Baerian formulation of embryogenesis has displaced that Haeckelian one: that is, the trend within embryogenesis is now taken as one of differentiation rather than the repetition of quasi-adult forms. In this respect, Von Baer's laws certainly contradict Haeckelian recapitulation. However, if Von Baer is interpreted to refer to a trend from more-commonly distributed to less-commonly distributed character states then such states may again be quasi-adult ones; and such character states are again being treated in an essentially Haeckelian way[33].

The temptation appears to be to treat character states as in some way autonomous, rather than as having their character defined by the adult state or by the context of the developmental system as a whole. This model of self-contained states is somewhat reminiscent of older models of genes as self-contained units strung out in a sequence. Whenever biologists treat character states as autonomous in this manner, the classical version of ontogeny appears. This step is most obvious in the writings of paleontologists and others, such as some cladists, who seek for familial connections between species. Developmental biologists are less concerned with such issues, and for them ontogeny as a sequence of states is more of a background or framework than a working model. Such versions of ontogeny, however, cannot but limit the scope of the associated research programme.

One contemporary biologist who has contributed both to paleontology and to developmental biology is Stephen Jay Gould. In *Ontogeny and Phylogeny* Gould demonstrated the importance of Haeckel's two-fold formulation in the history of biology, and indicated some of the deficiencies in Haeckel's phylogeny. These deficiencies in the Haeckelian view of evolutionary ancestry have now been discussed by Gould in a number of publications. As Gould shows, Haeckelian phylogeny typifies those aspects of 19th century biology which are being (slowly) overturned in the present century. It sees evolutionary history in terms of a simple pattern of ascent, through progressively 'higher' or more advanced species-types, up to humans at the top. In this respect the picture is that of Lamarck also. As Gould shows, the phylogeny picture relies on the notion of inevitable progress in the products of the natural world. In his latest book *Wonderful life*[34] Gould illustrates this with a number of pictorial versions of the 'ascent of man' - sequences of pre-human and pre-historic humans, each taller and more erect than the last, with facial angles gradually flattening and so on. For Gould, this human-centred view is simply vanity. There are no grounds for treating *Homo sapiens* as the inevitable and (in principle) predictable outcome of the evolutionary process. Evolution is not a ladder, and if it is a tree it is not one with a single trunk (Gould himself prefers the image of a bush).

Gould's criticisms of the progressivist view of phylogeny are entirely valid. They leave the term phylogeny empty. Since the work of Gould and of those he cites, if not since the demise of recapitulationism, phylogeny can only mean 'that sequence of historical occurrences which gave rise to such-and-such a modern species'. The word phylogeny is shorthand for reference to whatever postulated biological forms through history (simply, 'the past') are pertinent to the final outcome observed today. It is a convenient term in this respect, but it carries no explanatory weight; innocuous if clearly understood, but still capable of misleading the naive.

Given the strength of Gould's critique of progressivist phylogeny, one might expect him to be wary of progressivist tendencies in related spheres. But such is not the case. Indeed, a careful reading of *Ontogeny and phylogeny* shows that progressive change is vital to his theory. The first half of this book describes the rise and fall of Haeckelian recapitulation. The second half describes Gould's own reinterpretation of a number of the phenomena which had been taken as cognate with recapitulation - neoteny[35] in particular - and presents his argument for a general theory of heterochrony[36] (yet another Haeckelian term).

Gould assumes that certain aspects of an animal's lifespan career are fixed in the genetic material. Increments in size or shape, and onset of sexual maturation, are the most widely discussed examples. The timing of such changes is treated by Gould as independent in its programming from the changes themselves, so that alterations in timing (heterochrony) are relatively easy and serve as a source of variation on which selection can act. Gould's heterochrony theory relies on the notion of progressive, ordered series of character states (referred to in the context of his clock model as 'developmental stages'). For Gould, such series are robust and their constituents are relatively autonomous. The speeding-up or slowing-down of these sequences is assumed to be empirically demonstrable. Such changes in 'rate' can, it is argued, give rise to real changes in developmental outcome; indeed, lifespan development (and adult morphology) simply is the outcome of given sequences at given rates. Analysis in such terms is taken to have considerable explanatory power; as Gould repeatedly emphasises, retarded development "of itself and apart from any morphological correlates or consequences, has been a factor of paramount importance in human evolution".[37]

Gould's strongest empirical evidence derives from so-called neoteny in human evolution. Neoteny, for Gould, is an undeniable and almost observable fact in comparing humans with chimpanzees (the latter being taken to represent an ancestor of the human). Gould remarks "that a general retardation characterises human evolution can scarcely be denied"[38]. Neoteny, for Gould, is an evolutionary process involving delayed increment in shape while increment in size, and onset of sexual maturation, proceed as normal. Hence, compared to a putative ancestor, a descendent will show some features in adulthood which characterise the juvenile of the ancestor. In Gould's terminology, the process of neoteny gives rise to the result of paedomorphosis (Gould claims that human evolution is also characterised by hypermorphosis, the delay in sexual maturation while increment in size and shape proceed as normal[39]).

Note that 'delay' is being used in two senses in this argument: increment in shape is delayed within the lifespan of the descendent individual compared to the lifespan of the ancestral individual; but also, the descendent species has supposedly emerged by the evolutionary delaying of increment in shape in the ancestral species. There is an ambiguity here, of an ontogenetic process and an evolutionary process. The ontogenetic process is itself an outcome of the evolutionary process, so that the ambiguity is between delay as an outcome and as a process (a kind of ambiguity Gould is sensitive to elsewhere in the case of the term 'adaptation'). Gould exploits this ambiguity in the same way as did classical accounts of neoteny, in that the direct comparison of the lifespans of contemporary living individuals is taken as evidence for a putative historical process linking the one to an ancestor of the other. Readers of *Ontogeny and phylogeny* are left in no doubt as to the explanatory status of the supposed retardation in human evolution (from "Human development has slowed down" on p.9 to "the slow progress of [man's] life course" - here quoting Bolk - on p.404).

In this familiar argument, a common model is being employed for the developmental (lifespan) and the evolutionary (speciation) situations: the model is a single sequence of states in both cases. One species changes into another by the modulated speeding-up or slowing-down of its ontogenetic programme. In view of its ability to stretch, and to stretch differentially along its length, the image of a piece of elastic is appropriate. The transformation of species is interpreted as a linear transformation - a model somewhat less sophisticated than that of D'Arcy Thompson (to whom *Ontogeny and phylogeny* is dedicated) but, perhaps, within somewhat the same tradition.

The slippage between 'phylogenetic' and 'ontogenetic' claims is important because each supports the other in Gould's argument. Eldredge and Cracraft[40] argue that the appeal to neoteny cannot be granted any validity in the absence of some independent corroboration. That is to say, pointing out apparent delay or advance relationships (heterochrony) between contemporary species can do no more than suggest possible ancestral processes. I would go further and suggest that the mere identification of delays or advances is itself illegitimate, since to identify a case of delay or advance is at once to posit a causal process. The description becomes an explanation. Related problems of circularity have been identified by Alberch[41] in Nelson's 'biogenetic law', again involving the appeal to ontogenetic sequence as a self-explanatory observation. My point in this context, apart from casting doubt on the facticity of neoteny, is to highlight the role played by a linear model of development in this argument. Without a linear model for evolution, whether of species or of individuals, the argument from neoteny makes no sense.

But such linear models - whether ladders or pieces of elastic - have been rejected by Gould in other contexts. His examination of the history of intelligence testing, *The mismeasure of man*[42], makes the strong point that linear models of human intelligence cannot be acceptable. Treating cranial capacity as precisely correlated with intellectual achievement, for example, was a linear and hierarchical picture, and one intimately connected with the general progressive evolutionism of the last century. It is therefore strange that, in his discussions of heterochrony in general, and neoteny in particular, Gould employs such a simplistic version of lifespan change. The individual's career consists of progress through a series of prescribed states.

At this point the analysis of Alberch should be outlined, since he has already identified a number of these problems. According to Alberch, the appeal to ontogeny as a sequence of states is ubiquitous in the life sciences, from systematists to heterochronists. This model of ontogeny requires the homologising of stages across ontogenies (that is, across species). Independent evidence for the validity of such homologies is however absent. More generally, there are problems with treating ontogeny as a causal sequence of events. Empirically, there are often no meaningful intermediary stages between identifiable starting-points and end-points. This is the case in pattern formation, for example in the stripes of the zebra. Alberch argues that the stage-sequence model "is not compatible with the dynamical perspective of ontogeny defended by modern developmental biologists, particularly those working on problems of pattern formation and morphogenesis".[43]

Alberch's general conclusion is that the stage-sequence model for ontogeny is misleading and is to be abandoned. It must be noted that Alberch stops short of abandoning the term ontogeny itself. The sense in which he retains the term, however, is a loose and general one, the same kind of sense as he retains for phylogeny (as in 'the field of ontogeny and phylogeny').

Alberch's criticisms of the stage-sequence model of ontogeny are based on empirical as well as conceptual analyses. He is able to show that alternative formulations about lifespan change are already available to developmental biologists. One alternative approach not cited by Alberch is that of Goodwin[44], who has argued for the notion of generative transformations in developmental change. There are thus several alternatives to the stage-sequence model, emphasising that the stage analysis has no necessary validity.

Another kind of alternative to the classical stage-sequence model is the notion of the 'developmental pathway' or rather the landscape of such pathways. This model is perhaps the contemporary orthodoxy (in preference to the single sequence). At least superficially, the notion of a landscape with branching pathways would appear tree-like. Certainly, the notion of developmental pathways deliberately rejects the single-track account of species-specific ontogeny. It replaces a single track with a landscape containing multiple tracks, with decision points (as well as some force represented by gravity) constraining the free movement of an individual organism. Waddington's model of the 'epigenetic landscape' is a well-known formulation[45].

The notion of developmental pathways therefore represents a more sophisticated version of ontogeny than the stage-sequence, but appears to retain some of its basic features. Possibilities for the lifecourse are laid out in advance. Just as being born as a member of species A rather than species B immediately selects a lifecourse in the classical account of ontogeny, key environmental triggers here switch the growing organism between fixed alternatives. Indeed, the classical account of a universe of species-specific ontogenies could itself be seen as a kind of landscape of pathways. Available versions of the developmental pathway retain the prescription of developmental stages, the pre-existence of ontogenies, that is fundamental to the classical model. Pathways or channels are there prior to the unfolding development of the individual organism. In this respect, the notion of a landscape of developmental pathways does not represent an advance over the older notion of a fixed single sequence of developmental stages[46]. Notions equivalent to developmental pathways have, on the whole, been eschewed in the larger evolutionary context. Transcendental trends (toward greater size and complexity for example) are no longer given much explanatory weight[47]. But notions apparently akin to transcendental order are still maintained in the context of the developing individual. Why should this be?

The distinction that would usually be made is between the uniqueness of the set of events in evolution at large - speciation and the careers of species - and the repeatability of the life-courses of individual organisms. Newly-born members of a given species, it is assumed, repeat a cycle of changes in common with their con-specific organisms. A litter of newborn animals, for example, may contain ten individuals each of whom is destined to go through the same sequence of changes through to adulthood. The distinction, then, is between evolution as applying to a unique collection of events (and hence unrepeatable in principle) and ontogenesis as applying to any member of a class. The notion of ontogeny as predictable change depends on the class-membership status of each individual organism. More specifically, it depends on treating species as natural kinds. Since the work of David Hull and Michael Ghiselin, however, such assumptions have come under increasing scrutiny[48]. The implications of this situation are considered in the next section.

3. THE HISTORICAL ENTITY AND THE LIFE-CAREER: AN ALTERNATIVE TO ONTOGENY

Hull and Ghiselin have argued the case for the notion of individuality to be applied to single species as well as the more conventional application to single organisms. This claim has allowed Hull to investigate the historical nature of biological individuals in general. Any species, argues Hull, can be considered only as an 'historical entity', not as some natural kind of which the exemplars manifest some common essence. The formal notion of species in a classificatory system (in its hierarchical relation to 'class', 'order' and so on) might be class-like in this sense, but the species identified as human or kakapo is not. The historical entity is a unique sequence of events which should not be considered as a weak or obscured instance of a natural kind:"Historical entities are spatio-temporally localised particulars that develop continuously through time while staying internally cohesive".[49]

Organisms are "paradigm historical entities. Hence, they can function as central subjects in historical narratives". Hull is in this sense affirming the foundational importance of historical explanation within the life sciences, as against (at least in this context) natural-science explanation such as the covering-law.

If Hull's argument is valid for the species-as-individual, then it must also be valid for the individual-as-individual. Indeed, Hull confirms that (for example) the career of Napoleon Bonaparte has to be considered an historical entity, not a law-governed instance. This is not to say that at certain times Napoleon may not have instanced certain laws; but the career was not itself an instance of a law. To some extent, Hull's argument consists of the generalisation to species of what we already hold about (conventional) individuals. At the same time however, Hull endorses the notion of predictability (in the form of developmental pathways) for some kinds of individual - namely, 'a certain form of marine invertebrate'. Every member of the species in question (the species thus being treated as a class) has its sex determined by whether or not it lands on a con-specific's proboscis as it settles to the bottom of the pond in early life.

The claim I wish to make here is that Hull's distinction is not a valid one; that if Napoleon Bonaparte is (or was) an historical entity, then so is each of these particular invertebrates (let us here name one 'Josephine', which tells us that she missed all the available probosces). The qualitative distinction between male human and lowly invertebrate, however intuitive, is hard to justify. If Napoleon is granted status as historical entity, then so presumably must each human person. This status cannot depend on human-ness, and so must also be granted to each animal (and plant, etc). Perhaps, however, the distinction is made on somewhat different grounds. Napoleon's life-career might be treated as the combined product of two kinds of process: some law-bound developmental changes (in embryo and infancy, perhaps) and some law-free historical changes. It might be argued for example that Napoleon's maleness was decided in a logically analogous, if less spectacular, fashion as was Josephine's femaleness. If so, the life-career of Josephine the invertebrate might be analysed in a similar fashion to that of Napoleon, except that the historical component might be deemed negligible in her case.

Such a dualistic procedure is familiar to students of human development. It is a commonplace presupposition that the early (and also possibly very late) 'stages' of human life are closely regulated by genetic factors, with social-cultural-historical factors becoming massively effective in between[50]. This model turns out to be a slight revision of the traditional maturational account of human development in its accordance of predominant status to biological laws. This eclectic or compromise position cannot be disproven here except by appeal to the arguments in previous sections against stage-sequence accounts of individual change. Certainly, the force of Hull's innovative claims would be largely dispelled if such a compromise position were adopted. My view is that the exciting possibilities of the historical-entity formulation should be fully explored before such a retreat (let us call it Napoleon's retreat) is made.

The force of Hull's argument is that the careers of individuals cannot be taken at face value as instances of general laws. Such a claim runs counter to the orthodoxy of developmental biology, as exemplified by Hull's invertebrate example. Without seeking to pre-judge a complex and contested issue in the philosophy of biology, the possibility should at least be entertained that this orthodoxy is indeed wrong. Any name-able individual would be an historical entity, and unique. General laws, including general laws of developmental change, would then be inappropriate ("there are as many historical narratives as there are historical entities: the task is endless"[51]). Ontogeny, as investigated in this paper, seems to presuppose general laws of development. A sequence of stages is not the only way of representing a developmental law, so that ontogeny in general, not just the stage-sequence in particular, is threatened by this analysis.

Developmental laws are incompatible with historical-entity status. But historical entities do change, indeed, "Nothing whatsoever can change unless it is an individual"[52]. It is not clear from the literature whether such changefulness can by itself define the historical entity, or whether some identifiable starting-point is required. Such a starting-point, in any event, would seem to be readily available in the case of the life-careers of individual animals[53]. Like species, they are born and they develop. Every different individual is however unique, and it becomes superfluous to invoke a genetic programme since the unique historical sequence cannot be programmed.

My proposal is that the recent arguments of Hull and Ghiselin support and extend the arguments I have outlined above against current usages of ontogeny. Ontogeny is the attempt to grasp uniformities in life-careers by means of a covering-law explanatory method. This method is wholly inappropriate.

CONCLUSIONS

Discussions of developmental issues in the life sciences have been haunted by the conjunction of 'ontogeny and phylogeny' for well over a century. The search continues for the true connection of these two phenomena. The term phylogeny is now rather little used, other than colloquially, except in the context of this tantalising conjunction. When used in an explanatory sense, 'phylogeny' has been clearly identified as being committed to a pre-Darwinian progressivism: the assumption of transcendental order and direction in the upward march of Life. More colloquially, and without explanatory intent, 'phylogeny' simply refers to an undefined collection of historical events related together by an end result. Some vestiges of the explanatory sense of 'phylogeny' no doubt linger in the conjunction of this term with 'ontogeny', but the force of the conjunction - the implication that some meaningful interrelations are indeed to be discovered - probably derives from the apparent explanatory power of the latter term. That is to say, even if the term phylogeny is used with some diffidence in the conjunction 'ontogeny and phylogeny', the term ontogeny appears unexceptionable. It has been the intent of this paper to generalise the diffidence to the term ontogeny as well.

What I propose is that the term ontogeny should retain only a colloquial useage, exactly equivalent to the present useage of 'phylogeny'. That such an equivalence is appropriate is suggested by the conceptual relatedness of the two terms, from Haeckel to Gould. To use 'ontogeny' with explanatory intent is at least to run the risk of lapsing into pre-Darwinian formulations of biological change. Used colloquially, the term ontogeny refers to the empirical collection of events making up known life careers. Certainly, there would appear to be regularities within these empirical collections: tadpoles regularly turn into frogs, and caterpillars into butterflies. I wish to insist, however, that nothing is gained for theoretical advance or for research by treating such regularities as instances of natural laws. No matter how regular in terms of empirical observation, these apparently robust developmental sequences cannot be considered necessary but only contingent. (What they are contingent on is explored by Susan Oyama in this volume). The term ontogeny connotes intrinsic lawfulness and pre-programming - even if it is environmental interactions that are effectively said to be pre-programmed. The term itself cannot be blamed for its usage, but my own conviction is that our thinking would be well served if diffidence were to lead to embarrassment.

Individual lives are not predictable. They are fragile and may be snuffed out at any moment. (Parents of young children are only too aware of this fragility). There is an uncertainty principle at work. Replication across individuals (such as siblings) is imperfect, just as replication within an individual ('growth') is imperfect. The notion of ontogeny treats these replications as imperfect only in fact whereas the imperfection is in principle. There is no ideal or transcendent essence of the life-path of any particular species, and the notion of ontogeny may be nothing but a dim memory of such an ideal. Even trees of life may give us too ordered an image. Perhaps the image should include the community of scientists for whom, and for whom alone, the trees exist. Like it or not, when we gaze at nature we are gazing at ourselves.

ACKNOWLEDGEMENTS

This paper has been greatly improved through the generous advice of Paul Griffiths and editorial reviewers, contributors to the PPEB Conference, and David Hull.

NOTES

[1] P. Alberch (1985) 'Problems with the Interpretation of Developmental Sequences', *Systematic Zoology* **34**, pp.46-58.

[2] D. Hull (1984) 'Historical Entities and Historical Narratives', *Minds, Machines and Evolution: Philosophical Studies*, C. Hookway (ed), Cambridge University Press.

[3] See W. Kessen (1990) *The Rise and Fall of Development*, Clark University Press for a discussion of the evolution-progress-development complex - "a triad almost as arrogant as 'For God, for Country, and for Yale'".

[4] However, see P.P.G. Bateson (1987) 'Biological Approaches to the Study of Behavioural Development', *International Journal of Behavioural Development* **10**, pp.1-22.

[5] J.R. Morss (1990) *The Biologising of Childhood: Developmental Psychology and the Darwinian Myth*, Lawrence Erlbaum Associates. Also see M. Ghiselin (1986) 'The Assimilation of Darwinism in Developmental Psychology', *Human Development* **29**, pp.12-22. Ghiselin notes (p.12) "The history of the assimilation of Darwinism has been the history of failure to assimilate Darwinism ... This is particularly true of psychology, including developmental psychology".

[6] S. Oyama (1989) 'Ontogeny and the Central Dogma: Do we need the Concept of Genetic Programming in Order to Have an Evolutionary Perspective?', M. Gunnar and E. Thelen (eds.), Systems and Development, *The Minnesota Symposia on Child Psychology* **22**, Lawrence Erlbaum Associates; also Oyama, this volume.

[7] For a discussion of diverse historical approaches to biology, see R.J. O'Hara (1988) 'Homage to Clio, or, Toward an Historical Philosophy for Evolutionary Biology', *Systematic Zoology* **37**, pp.142-155.

[8] This section is based on my book (Morss, op. cit.) to which reference should be made for documentation on all points made here, unless otherwise noted. The general importance of the wider, non-Darwinian contexts within which evolutionary ideas were expressed has been emphasised by Peter Bowler in a number of publications. See, in particular, P. Bowler (1983) *The Eclipse of Darwinism: Anti-Darwinian Evolution Theories in the Decades around 1900*, Johns Hopkins University Press; P. Bowler (1988) *The Non-Darwinian Revolution: Reinterpreting a Historical Myth*, Johns Hopkins University Press.

[9] It would be a serious mistake to identify Darwinism with a 'developmental' approach to biology. As pointed out by Sober (1984) *The Nature of Selection*, MIT Press, the term 'development' (in an 'evolutionary' context) might itself seem to connote a sequence of progressive stages - for example, in the style of Lamarck. If so Darwinism, with its stress on variation and stasis, would be inherently anti-developmental. Similarly, Bowler (1988, op. cit.) argues that 19th Century biology as a whole should be described as 'developmental'. Bowler suggests that the term 'evolution' was extrapolated to the history of speciation from a more specifically ontogenetic context. The effects of such an extrapolation are still discerned by some observers; O'Hara, for example, stating that "Biologists must free themselves from the ontogenetic view of evolution" (op. cit., 153). The conceptual relationship between what Haeckel called phylogeny and what he called ontogeny are complex and bidirectional. It is clear, however, that the commitment to a developmental evolutionism is a pre-Darwinian one.

[10] Haeckel defined ontogeny as 'germ-history, or the history of the evolution of the individual' and phylogeny as 'tribal history or the palaeontological history of evolution': E. Haeckel (1879) *The Evolution of Man*, Kegan Paul, Vol. II p.460.

[11] W. Preyer (1896); cited in G. Eckardt (1985) 'Preyer's Road to Child Psychology' in G. Eckardt, W. Bringmann and L. Sprung (eds) *Contributions to a History of Developmental Psychology*, Mouton.

[12] W. Preyer (1869; 1897); cited in Morss, op. cit., 28.

[13] R. Richards (1977) 'Lloyd Morgan's Theory of Instinct: From Darwinism to Neo-Darwiniam', *Journal for the History of the Behavioural Sciences* **13**, pp.12-32; cf p.15.

[14] Darwin's 'Biographical Sketch of an Infant' is discussed in B. Bradley (1989) *Visions of Infancy*, Polity Press.

[15] Bowler (1983) op. cit.; (1988) op. cit.

[16] For a general discussion of the widespread impact of evolutionist thinking, see S.J. Gould (1977) *Ontogeny and Phylogeny*, Belknap Press.

[17] G.S. Hall (1904) *Adolescence: Its Psychology and its Relations to Physiology, Anthropology, Sociology, Sex, Crime, Religion and Education*, Appleton.

[18] J. Chisholm (1988) 'Toward a Developmental Evolutionary Ecology of Humans', K. MacDonald (ed) *Sociobiological Perspectives on Human Development*, Springer-Verlag.

[19] F. Sulloway (1979) *Freud, Biologist of the Mind*, Basic Books; also S.J. Gould (1987) 'Freud's Phylogenetic Fantasy', *Natural History*, December, pp.10-20.

[20] Ghiselin, op. cit., discusses Piaget's "failure to understand evolution" ("Piaget's real intellectual ancestor was Spencer"); Vidal et al. suggest that Piaget misunderstood the significance of Mendelism: F. Vidal, M. Buscaglia, and J. Vonéche (1983) 'Darwinism and Developmental Psychology', *Journal for the History of the Behavioural Sciences* **19**, pp.81-94.

[21] The title is more innocuous in English translation: J. Piaget (1979) *Behaviour and Evolution*, Routledge and Kegan Paul.

[22] As well as Morss op. cit., see the more detailed account of this 'three-

mountains' task in Morss (1987) 'The Construction of Perspectives: Piaget's Alternative to Spatial Egocentrism', *International Journal of Behavioural Development* **10**, pp.263-279. Piaget re-reported the study (by Edith Meyer) on several occasions, each time giving greater prominence to certain ('egocentric') features of the results.

[23] Lorenz' debt to Bölsche is discussed by T. Kalikow (1983) 'Konrad Lorenz' Ethological Theory: Explanation and Ideology, 1938-1943', *Journal of the History of Biology* **16**, pp.39-73. Lorenz' early writings on evolution are reprinted in R. Evans (1975) *Konrad Lorenz: The Man and his Ideas*, Dutton.

[24] Detailed substantiation of this claim is given in Morss (1990), op. cit., 199-204.

[25] W. Hodos and C. Campbell (1969) 'Scala Naturae: Why there is no Theory in Comparative Psychology', *Psychological Review* **76**, pp.337-350.

[26] Oyama, this volume.

[27] Although the term ontogeny has not been extensively criticised within developmental psychology, it is important to note that stage theories of development have been. For example, the Piaget-Kohlberg stage theory of moral development has been criticised by R. Harré (1983) *Personal Being*, Blackwell.

[28] D. Rosen (1984) 'Hierarchies and History', J. Pollard (ed) *Evolutionary Theory: Paths into the Future*, Wiley.

[29] J. Eldredge and J. Cracraft (1980) *Phylogenetic Patterns and the Evolutionary Process*, Columbia UP.

[30] R. Fortey and R. Jefferies (1982) 'Fossils and Phylogeny: A Compromise Approach', K. Joysey and A. Friday (eds) *Problems of Phylogenetic Reconstruction*, Academic Press.

[31] P. Janvier (1984) 'Cladistics: Theory, Purpose and Evolutionary Implications', in Pollard, op. cit.

[32] Alberch, op. cit., 56.

[33] On Von Baer, see Eldredge and Cracraft, op. cit., 60; also, Alberch's comment on Nelson's version of the 'biogenetic law'.

[34] S.J. Gould (1990) *Wonderful Life*, Hutchinson.

[35] Neoteny is defined (Gould, 1977, op. cit., 483) as 'Paedomorphosis (retention of formerly juvenile characters by adult descendents) produced by retardation of somatic development'. The term was originated by Kollmann in 1885.

[36] Heterochrony is defined by Haeckel (op. cit., Vol. I, 12) as 'a kenogenetic vitiation of the original, palingenetic incidents of evolution' by displacement in time of the phenomena, resulting from adaptation to changed conditions. Gould (op. cit., 482) adopts De Beer's definition of 'phyletic change in the onset or timing of development'.

[37] Gould, op. cit., 399.

[38] Ibid, 366.

[39] A logical point should perhaps be made here. Gould's argument is that neoteny and hypermorphosis have been of special significance for human evolution. Gould, op. cit., defines these two processes as two of the six permutations among three factors (increase in size, increase in shape, onset of sexual maturation) one of the three being either slowed down or speeded up in each case. (3 factors multiplied by 2 types of rate change = 6 possible kinds of heterochrony). In neoteny, the second of these factors is retarded; in hypermorphosis the third of these is retarded. Now the combined effect of neoteny and hypermorphosis together - second and third factors both retarded - is, logically, equivalent to the acceleration of the first factor alone. This process is

giantism - a simpler but perhaps less interesting model for human evolution.

[40] Eldredge and Cracraft, op. cit., 62.

[41] Alberch, op. cit.

[42] S.J. Gould (1981) *The Mismeasure of Man*, Penguin Books.

[43] Alberch, op. cit.

[44] B. Goodwin (1984) 'Changing from an Evolutionary to a Generative Paradigm in Biology', in Pollard, op. cit.

[45] C. Waddington (1957) *The Strategy of the Genes*, Allen and Unwin.

[46] Gould's developmental pathways tend to reduce to a single sequence in rather the same way as tree-like representations of 'phylogeny' tend to reduce to single trunks: see R. O'Hara 'Telling the Tree: Narrative Representation and the Study of Evolutionary History' (MS submitted to Systematic Zoology, 1991).

[47] Transcendental trends are still to be found in the literature however. Trends to increased size and complexity are discussed by Bonner (1988) *The Evolution of Complexity by Means of Natural Selection*, Princeton University Press. Butler (in Joysey and Friday, op. cit.) describes 'reversed evolution' in the history of dentition, but only with respect to single characters; "Such reversals are secondary consequences of progressive evolution of a wider functional system to which the characters are subordinate". This 'saving' of progressivism is a familiar technique in developmental psychology. Stephen Jay Gould remarked ('Living Treasures' lecture, University of Otago, 1990) that evolution sometimes "goes into a different channel" as a result of a certain environmental event. It is difficult to conceive of a 'channel' except as in some sense pre-existing the going-into. For an exploration of an interactive interpretation of developmental pathways, see Oyama, this volume.

[48] Hull, op. cit.; M. Ghiselin (1988) 'Species Individuality has no Necessary Connection with Evolutionary Gradualism', *Systematic Zoology* **37**, pp.66-67.

[49] Hull, op. cit., 17.

[50] Morss (1990), op. cit., 205.

[51] Hull, op. cit., 33.

[52] Ghiselin (1988) op. cit., 66.

[53] The oral presentation of this paper carried the title 'Species of Origin' and it would seem that considerable explanatory weight is placed on the reality of an origin in the historical-entity formulation, as it has been in more traditional accounts of development. Whether this reliance on origins is either necessary or desirable is as yet unclear.

John Morss,
Department of Education,
University of Otago.

INDEX

AUSTRALIAN STUDIES
IN HISTORY AND PHILOSOPHY OF SCIENCE

General Editor:
R. W. Home, *University of Melbourne*

Publications:

1. R. McLaughlin (ed.): *What? Where? When? Why?* Essays on Induction, Space and Time, Explanation. Inspired by the Work of Wesley C. Salmon. 1982 ISBN 90-277-1337-5
2. D. Oldroyd and I. Langham (eds.): *The Wider Domain of Evolutionary Thought.* 1983 ISBN 90-277-1477-0
3. R. W. Home (ed.): *Science under Scrutinity.* The Place of History and Philosophy of Science. 1983 ISBN 90-277-1602-1
4. J. A. Schuster and R. R. Yeo (eds.): *The Politics and Rhetoric of Scientific Method.* Historical Studies. 1986 ISBN 90-277-2152-1
5. J. Forge (ed.): *Measurement, Realism and Objectivity.* Essays on Measurement in the Social and Physical Science. 1987 ISBN 90-277-2542-X
6. R. Nola (ed.): *Relativism and Realism in Science.* 1988 ISBN 90-277-2647-7
7. P. Slezak and W. R. Albury (eds.): *Computers, Brains and Minds.* Essays in Cognitive Science. 1989 ISBN 90-277-2759-7
8. H. E. Le Grand (ed.): *Experimental Inquiries.* Historical, Philosophical and Social Studies of Experimentation in Science. 1990 ISBN 0-7923-0790-9
9. R. W. Home and S. G. Kohlstedt (eds.): *International Science and National Scientific Identity.* Australia between Britain and America. 1991 ISBN 0-7923-0938-3
10. S. Gaukroger (ed.): *The Uses of Antiquity.* The Scientific Revolution and the Classical Tradition. 1991 ISBN 0-7923-1130-2
11. P. Griffiths (ed.): *Trees of Life.* Essays in Philosophy of Biology. 1992 ISBN 0-7923-1709-2

KLUWER ACADEMIC PUBLISHERS – DORDRECHT / BOSTON / LONDON

ERRATA

GRIFFITHS (ed.): TREES OF LIFE
ISBN 0-7923-1709-2 1992
Australasian Studies in History and Philosophy of Science
Volume 11

SUSAN OYAMA
ONTOGENY AND PHYLOGENY; A CASE OF METARECAPITULATION?

p. 214, fig. 2, reverse 'Phylogeny' and 'Ontogeny' in margin and correct 'nurture'
line 2, 'causal' in lower case

p. 229, note 5, line 1: 'Oyama (1991)'
note 3, 5, 6, 'Ho and S.W. Fox'

p. 230, note 6, line 3 'pp. 21-27

p. 231, note 10, line 12: 'pp. 1-44'
note 11, line 6: 'pp. 1-13'

p. 232, note 15, line 2: 'pp. 1-68'
note 16, lines 7-8; 'without recourse to contingencies' should be in italics

p. 233, note 23, S. Oyama (1991) 'Bodies and Minds', Journal of Social Issues, XL (3), pp. 27-42.
note 26, delete second word

p. 237, note 43, insert at beginning of note: R.C. Lewontin (1982) 'Organism and Environment', Learning, Development, and Culture, H.C. Plotkin (ed.), Wiley

MIX
Papier aus verantwortungsvollen Quellen
Paper from responsible sources
FSC® C105338

If you have any concerns about our products,
you can contact us on
ProductSafety@springernature.com

In case Publisher is established outside the EU,
the EU authorized representative is:
Springer Nature Customer Service Center GmbH
Europaplatz 3, 69115 Heidelberg, Germany

Printed by Libri Plureos GmbH
in Hamburg, Germany